The Future of Ageing in Europe

Alan Walker
Editor

The Future of Ageing in Europe

Making an Asset of Longevity

palgrave
macmillan

Editor
Alan Walker
Department of Sociological Studies
University of Sheffield
Sheffield, UK

ISBN 978-981-13-1416-2 ISBN 978-981-13-1417-9 (eBook)
https://doi.org/10.1007/978-981-13-1417-9

Library of Congress Control Number: 2018955833

Cover illustration: Seb Oliver

This Palgrave Macmillan imprint is published by the registered company Springer Nature Singapore Pte Ltd.
The registered company address is: 152 Beach Road, #21-01/04 Gateway East, Singapore 189721, Singapore

To Mabel and Bruce

Preface

This book marks the completion of the major European project Mobilising the Potential of Active Ageing in Europe (MOPACT), which engaged 32 partner institutions, mainly research institutes and universities, in 13 countries. It brought together a large group of leading European experts on the various dimensions of population ageing and its policy implications, together with a talented cohort of doctoral and postdoctoral early career researchers. Their considerable expertise and full dedication to the achievement of the project's objectives are apparent in this book. Of course it is impossible to include every finding or analysis from a complex research project spanning five years, so readers are invited to visit the MOPACT website http://mopact.group.shef.ac.uk/ to access further information. There you will find extensive resources on active ageing in Europe, including many examples of good practice, all of the project's scientific outputs, and lay summaries of the key elements of the research.

In offering thanks, the first of these must go to the European Union's Horizon 2020 research and innovation programme (FP7-SSH.2012-1/ No 320333) without which this research would not have been possible (Brexiteers please note). Several project officers from the European Commission's DG Research and Innovation worked with us over the life of the project, all of whom were extremely supportive and none more so than Georgios Papanagnou.

Second, my thanks go to all of the researchers who took part in MOPACT: thank you for your excellent work and support. Thanks also to the many stakeholders who kept the research focused on real-world needs. There are too many to mention by name but our partner AGE Platform Europe must be included, especially Ilenia Gheno, Philippe Seidel Leroy, Alice Sinigaglia, Nhu Tram and Anne-Sophie Parent. Also, various valued colleagues ensured a close connection was maintained with the issues facing the Central and Eastern European Member States: the staff at PRAXIS in Estonia, especially Reelika Leetmaa and Andres Võrk; the Romanian National Research Institute for Labour and Social Protection (INCSMPS), especially Speranta Pirciog and Mihaela Ghenta; and CASE in Poland, especially Elena Jarocinska, Izabela Styczyńska and Aart Jan Riekhoff. Warm thanks to them. Also, many thanks to Jørgen Mortensen who kept a watchful eye on the project and was very helpful at the application stage. Mikkel Barslund, Kai Leichsenring, Tony Maltby, Gerd Naegele, Frank van der Duyn Schouten and Asghar Zaidi also made important contributions at the application stage.

Third, no project of the size and scope of MOPACT can be managed by one person. In this case, two very special people were vital to its operation and successful conclusion: Juliet Craig and Sarah Howson, both of whom are MOPACT stars. Juliet was also absolutely crucial to the success of the original application. Over the life of the project, various other administrative staff worked with them on the operational front, sincere thanks to them: Peter Adcock, Dina Al-Huraibi, Adele Blinston, Stephanie Gordon, David Neary, Fiona Oliver and, for constant financial expertise, Anya Rayner.

Fourth, in the completion of this volume I have been lucky to receive extraordinary backup from the indefatigable and very talented Vanessa Rodgers. Also important have been the generous expert inputs of Asghar Zaidi, especially with regard to Chaps. 1 and 11 and Dan Holman with respect to Chap. 1. Sincerest possible thanks to them and everyone else mentioned, and abject apologies for any oversights.

Finally, this book is dedicated to my grandchildren Mabel and Bruce in the hope that their ageing will be enhanced by the adoption of the key policy recommendations contained herein. We live in hope.

Sheffield, UK Alan Walker

Contents

List of Figures

List of Tables

1

Introduction

Alan Walker

Population ageing is acknowledged by governments globally as one of the grand policy challenges. This is especially the case in Europe because it is the most aged continent in terms of the proportion of its population aged 65 and over. Thus, this issue has been high on the research and policy agendas for several decades (Walker and Maltby 1997), and the European Union (EU) has played a prominent role both in drawing the attention of Member State governments to the implications of population ageing and in searching for optimum policy responses. The origins of the EU's coordinating role in this field can be traced back to the European Year of Older People in 1993, which represented the first proclamation by Europe of the idea of a new active and participative discourse on ageing. It has been followed up in the intervening period by numerous demographic and policy reports and, while these have not always been consistent in their approach and have often been overly

A. Walker (✉)
Department of Sociological Studies, University of Sheffield, Sheffield, UK
e-mail: a.c.walker@sheffield.ac.uk; https://www.sheffield.ac.uk/socstudies/staff/staff-profiles/walker

A. Walker (ed.), *The Future of Ageing in Europe*,
https://doi.org/10.1007/978-981-13-1417-9_1

economistic (Walker 2009; Walker and Foster 2013), the European Commission has consistently pursued the goal of promoting greater participation among older people, even though this has often been rather myopically focused on the labour market. This was definitely not the case with regard to the very specific goal of HORIZON 2020, established in 2010, and which provided the spur for the research upon which this book is based. Within the innovation framework of HORIZON 2020 the European Commission (EC) established the European Innovation Partnership on Active and Health Ageing (EIPAHA) with the ambitious aim of raising healthy life expectancy (HLE) across the EU by an average of two years by 2020. This bold target provided the context for the EC's Research Directorate to issue a call for proposals to investigate how increasing longevity may be an 'asset for Europe'. The successful application for funding, entitled Mobilising the Potential of Active Ageing in Europe (MOPACT), is one of the largest multi-disciplinary projects on ageing ever mounted in Europe. Its scale and disciplinary spread warrant its status here as the centrepiece of this volume, as the next section demonstrates.

MOPACT

The MOPACT project comprised 32 partners in 13 countries, including more than 100 researchers, and took place between 2012 and 2017. It sought to concentrate the highest possible quality of scientific analyses on the key policy issues in ageing and, further, to develop innovative policies and practice aimed at assisting public authorities and other key actors, at all levels in Europe, to transform approaches to this grand challenge. This huge research and development effort was focused specifically on five key dimensions of the social and economic impact of ageing:

- Economic and financial consequences of ageing
- Societal structures, civil society and cohesion
- Social support, long-term care (LTC) and quality of life
- The built and technological environment
- Health, well-being, biogerontology and frailty

These five key themes were tackled in nine scientific thematic clusters or work packages, which comprise the main chapters of this book:

- Realising active ageing (Chap. 2)
- Economic consequences of ageing (Chap. 3)
- Pension systems, savings and financial education (Chap. 4)
- Extending working lives (Chap. 5)
- Health and well-being (Chap. 6)
- Biogerontology (Chap. 7)
- Built and technological environments (Chap. 8)
- Social support and LTC (Chap. 9)
- Enhancing active citizenship (Chap. 10)

The first of these work packages acted as the core for the whole project, not only pursuing its own distinct research agenda but also continuously synthesising the results from the work of the other thematic clusters. This process helped to avoid the tendency of such large multi-dimensional projects to operate along parallel lines rather than as an integrated whole. This integration was reinforced by regular work package 'leaders' and periodic whole team meetings. In particular, the inclusion of biogerontologists within a policy-orientated project is novel and a special case had to be made for it in the application process. The validity of this cause is demonstrated in the depth this perspective added to our understanding of the potential for life course interventions to improve well-being (Chap. 7 and *passim*).

Given the explicit policy and practice focus of MOPACT, there also had to be a high level of integration of the various stakeholders in this field. We achieved this by, first, ensuring that all work packages engaged with relevant stakeholders and, second, by adopting the model of stakeholder forums pioneered by ERA-AGE (http://era-age.group.shef.ac.uk/). The latter consisted of a series of feedback loops wherein researchers presented their work to a forum of policymakers, practitioners and product producers and received constructive suggestions on its relevance and how it might be developed further. Three such forums were held (Fig. 1.1) and, because of the highly uneven nature of the development of policies on ageing between Member States, with the Central and Eastern European

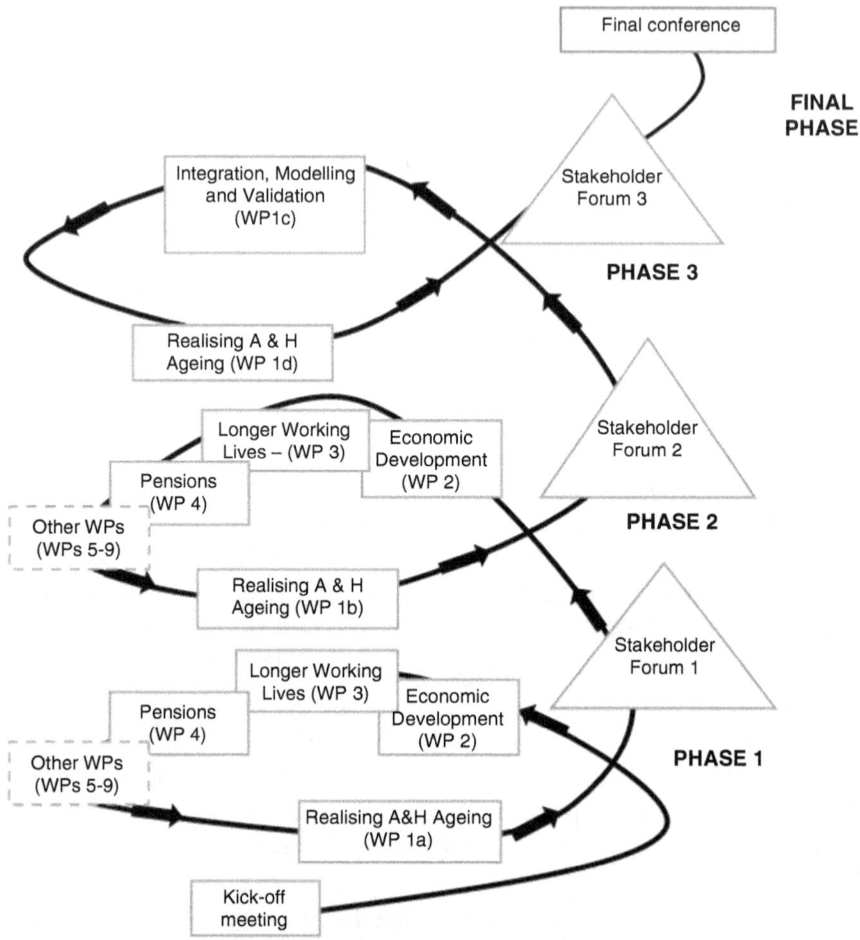

Fig. 1.1 The MOPACT consultative architecture

countries invariably lagging behind their Western counterparts (see Chap. 2 and *passim*), two of these forums were staged there. More than 100 stakeholders were involved in the forums.

To complete the sketch of how MOPACT was organised, there were also two work packages concerned with management and dissemination. In all there were more than 50 reports and other outputs from the project, including 12 lay summaries of the main findings from each of the work packages. This book represents a synthesis of that huge body of research. The MOPACT website contains further details of the activities of the project and its outputs: http://mopact.group.shef.ac.uk/.

Centrality of Active Ageing

Why was active ageing placed at the heart of MOPACT? First of all we followed the prescription of the Road Map for European ageing research (Futurage 2011) because it represented the most comprehensive consultation on this topic ever undertaken in Europe, spanning all scientific disciplines relevant to ageing and adopting a comprehensive approach. On the basis of this extensive groundwork, the Road Map argued that active ageing across the life course should be the keystone of all research in this field. The Road Map made a compelling case for this centrality of active ageing. It is a helpful umbrella term to encompass various quality of life essentials such as labour market participation, active contribution to domestic labour (caring, housework), active participation in community life and active leisure. It is also valuable in synthesising strands of research on ageing and developmental science which traditionally have not had much in common. For example, the need to combine research able to drive social policy or cultural investments with that concerning the individual level of ageing, such as with regard to health, cognitive functioning and motivation. In social gerontology too, the concept of active ageing links the macro, meso and micro aspects of ageing and ageing research. Furthermore, active ageing requires a social-ecological view of ageing (Lawton 1980; Wahl and Iwarsson 2007). Different levels of evidence-based policy action (macro), community and neighbourhood arrangements (meso) and individual intervention (micro) must go hand in hand in order for active ageing to be realised, as discussed in more detail in Chap. 2. Because of the interplay between active ageing and these different social and environmental dimensions our approach to it must also be multi-disciplinary—no single discipline can provide the evidence necessary to support it. This, in turn, means that the concept of active ageing has the power to connect disparate disciplines—sociology, social gerontology, psychology, engineering, economics, medical sciences and biogerontology—and build bridges between them. This proved invaluable in the MOPACT project because of its multi-disciplinary nature and the need to integrate the five major themes it focused on.

Secondly, we were keen to build on the pioneering work undertaken by the EC/UNECE's Active Ageing Index (AAI) Project (Zaidi and Stanton 2015; Zaidi et al. 2017), which began as one of the activities of the 2012 European Year for Active Ageing and Solidarity Between Generations. As explained in greater detail below and in Chap. 2, the AAI provided a helpful starting point for the development of a deeper understanding of the strategies required to promote active ageing and how these vary across Europe.

Thirdly, while there were compelling scientific reasons for the centrality of active ageing, the political ones were no less so. The European Years of Older People (1993) and Active Ageing (2012) have been mentioned already. In between them there were several elaborations of what a European policy on active ageing might consist of; the most comprehensive and clearly envisioned of them was the one prepared for the 1999 UN Year of Older People (EC 1999). At the global level, the World Health Organisation (WHO) has advanced a multi-dimensional concept of active ageing:

> The process of optimising opportunities in health, participation and security in order to enhance quality of life as people age. Active ageing applies to both individuals and groups. It allows people to realise their potential for physical, social and mental well-being throughout their lives and to participate in society according to their needs, desires and capacities, while providing them with adequate protection, security and care when they require assistance. (WHO 2002)

This important conceptualisation emphasised the case for ensuring that active ageing is not a narrow idea, focused only on the labour market, but a broad one that embraces all of the different factors that contribute to personal well-being. Specifically, it placed participation at the centre of consideration rather than employment. In addition, it stressed the initial importance of a life course perspective: in order to *prevent* some of the negative aspects of growing old, it is essential to influence individual behaviour and its policy context at earlier stages of the life span. The WHO's approach to active ageing also reinforced the growth of the positive discourse on older people as active participants in society stretching back to 1993. The priority of active ageing was adopted by the UN's Madrid International Action Plan on Ageing, in 2002, along with the principle of older people's right to participate (UN 2002; Sidorenko and Walker 2004).

It was for these scientific and political reasons that we adopted the idea of active ageing as an organising concept within MOPACT. But we did not do so uncritically. First of all, following the European ageing research Road Map, we adopted a broad view of 'activity' to include all meaningful pursuits, mental, as well as physical, that contribute to well-being. This is to ensure that the common reduction to working life would be prevented, within the project at least. Secondly, we enforced a whole life course approach to ensure specifically that the oldest old were not excluded, as is so often the case in discussions of active ageing. This novel stand is demonstrated strongly in Chap. 9, where the concept is operationalised with regard to LTC. Thirdly, we added a multi-dimensional division of labour to the WHO definition, to underline the fact that active ageing depends on a wide range of different actors and cannot simply be imposed top-down by policymakers. Thus, active ageing is a comprehensive strategy to maximise participation and well-being as people age. It should operate simultaneously at the individual (lifestyle), organisational (management) and societal (policy) levels, and at all stages of the life course.

Nor were we uncritical of the European politics of active ageing. As indicated there has been at least one bold, visionary statement by the EC on what active ageing could be. Too often, however, policy is reduced, only or primarily, to the goal of extending working lives, which is focused on older workers where, paradoxically, a longer term life span orientation is required (Walker 2009 and Chap. 5). This mis-direction of the policy focus is taken up again in Chap. 11.

As well as the centrality of active ageing we also derived from the European ageing research Road Map eight basic principles:

- An holistic person-environment approach to ageing
- A life course perspective

- Multi-disciplinarity
- User and stakeholder engagement
- Knowledge exchange and impact
- Diversity: structural, cross-cultural and cross-national
- Capacity building in research
- Maximising technological innovation

These basic principles were observed in the work of MOPACT and its organisation, and are evident in the following chapters.

Social Innovation

The concept of social innovation was also adopted by MOPACT, although it did not occupy the same central position as active ageing. It is an idea that has achieved very wide currency in Europe, hence its adoption by HORIZON 2020 and the EIPAHA. In practice, however, it lacks a clear definition and is '… a term that almost everyone likes, but nobody is sure of what it means' (Pol and Ville 2009, p. 881). Drawing heavily on the work of Murray et al. (2010) the EC has defined it as

> … the development and implementation of new ideas (products, services, models) to meet social needs and create new social relationships or collaborations. It represents new responses to pressing social demands, which affect the process of social interactions. It is aimed at improving human well-being. Social innovations are innovations that are social in both their ends and their means. They are innovations that are not only good for society but also enhance individuals' capacity to act. (European Commission 2013, p. 6)

Unfortunately, this definition arrived after the MOPACT project had commenced and, in any case, this one is rather cumbersome and not specifically related to the ageing field. Fortunately, though our sister project Social Innovations Promoting Active and Healthy Ageing (INNOVAGE http://www.innovage.group.shef.ac.uk/) had faced the same problem and had developed its own definition:

> Social innovations are ideas, products, services or models that are new, or applied in new contexts, and which are designed to improve the well-being and quality of life of people as they age. (Neary et al. 2015, p. 6)

This streamline definition was adopted because it had already been operationalised successfully by INNOVAGE and was fit for purpose. It was also tailor-made to reflect the multi-dimensional approach of MOPACT. Thus, social innovations might range from micro-level changes in relations between individuals and groups to macro-level policy changes that may even result in systemic change. They may be both 'top-down' official initiatives and 'bottom-up' civil society ones. This definition was not imposed on the whole project by the coordinator and therefore, there was scope for some variation in approach according to the focus of the specific work packages. For example, for the purposes of the research on extending working lives (Chap. 5), Naegele and Bauknecht (2014, p. 16) developed this definition:

> … an intentional, purposeful new configuration of social practices realised by a certain group of stakeholders whose objective is to solve socio-political problems or satisfy needs better than it would be possible on the basis of established practices.

The core work package on realising active ageing applied the INNOVAGE definition and also adopted the methodology developed by that project, in conjunction with the Young Foundation, for classifying social innovations.

The objective was to conduct a global web-search for social innovations in the field of active ageing which, if suitable, would become exemplars for publication on the MOPACT website, which would then operate as a reference point for those seeking to develop their own projects. An extensive survey was undertaken and a special effort was made to find innovations in the Central and Eastern European so-called new Member States. To be included in our long list of potential exemplars, these needed to demonstrate robust potential and stakeholder support. Then these were processed through an evaluation based on the balanced scorecard developed by INNOVAGE but adopted for MOPACT's purposes. There were four components to the scorecard—social and economic impact, sustainability, tolerance, and implementation—each of which had two indicators. With a maximum score of five on each of the four components (1 = low, weak, poor 5 = very good, highly likely, fully achieved), the maximum score for each project was 20. Only those that achieved a minimum threshold of 12 in this evaluation process were uploaded to the MOPACT web portal as exemplars of active ageing, 150 of them. These were classified under the headings used in the AAI (below and Chap. 2) and can be accessed via the MOPACT website.

Ageing Europe

It has been noted already that Europe is the oldest region in the world in terms of the proportion of its population who are classed as being 'old' (usually 65 and over), which stands at 16 per cent. The ageing of Europe's population is a long-term trend which, along with similar trends in other developed countries, can be traced back to the mid-nineteenth century (EC 2004; Oeppen and Vaupel 2002). It results from two demographic transitions: on the one hand death rates at older ages are falling, which increases longevity, and on the other hand, fertility is declining, which leads to a rise in the average age of the population. Inward migration could offset the ageing trend, providing that the average age of the migrants was low, which it is usually, but the overall numbers have not been sufficient to do so (outward migration is important at national level as it may leave care and social contribution gaps). Thus, the average age of the European population is rising and is projected to continue to do so. These demographic changes are summarised in the population pyramid below.

Although there are annual fluctuations and differences between countries, the long-term demographic trend has seen a sharp decline in fertility rates following the post-war 'baby booms', which saw a peak above 2.5 children per woman of child bearing age in the second half of the 1960s. Fertility rates fell below the natural replacement rate of 2.1 in the late 1960s in Czechia, Denmark, Finland, Germany, Hungary, Latvia, Luxembourg and Sweden. These forerunner countries were followed, in the 1970s, by Austria, Belgium, France, Italy, the Netherlands and the

UK. It was not until the next decade that similar falls in fertility rates occurred in Greece, Malta, Poland, Portugal, Slovakia and Spain. Ireland was the last Member State of the EU to join the low fertility club, in 2000. Now no EU country has a total fertility rate at or above the replacement rate. France and Ireland come closest at 2.0, compared to the EU28 average of 1.59, but their rates are projected to decline further to 1.98 by 2060. Over the same period the EU28 average is also projected to decline further to 1.76 (EC 2014, p. 8).

In contrast to fertility rates, life expectancy has been increasing steadily, if unevenly, in Europe. Also the historic difference in life expectancy between men and women has been declining since 1990 as a result of the relative improvement in male life expectancy. As with fertility rate changes, there are significant differences between EU countries in life expectancy gains. Taking life expectancy at birth, between 1960 and 2012, women gained 11 years or more in Finland, France, Germany, Ireland, Italy, Luxembourg, Malta, Portugal, Slovenia and Spain. In the same period smaller increases, of eight years or less, were recorded in Czechia, Denmark, Latvia, the Netherlands and Slovakia. Gains in life expectancy, 1960–2012, for men were 11 years or more in Austria, Belgium, France, Finland, Germany, Luxembourg, Malta, Portugal, Slovenia, Spain and the UK. Increases of eight years or less took place in Bulgaria, Croatia, Czechia, Denmark, Estonia, Hungary, Latvia, Lithuania, the Netherlands, Poland and Slovakia (EC 2014, p. 10). It is the rise in life expectancy that is driving the most significant demographic trend shown by Fig. 1.2 which is the increase in the very elderly or 'old old' age groups, particularly among women. This advanced ageing is a major issue for both society and policymakers, for example, in the LTC field, and therefore is a recurring one in this book.

Looking forward, life expectancy at birth for men is expected to increase by 7.2 years between 2013 and 2060—from 77.6 years to 84.7 years. For women the increase is expected to be 6 years—from 83.1 years to 89.1 years. The largest rises in life expectancy for both men and women are projected to take place in the countries with the lowest life expectancies in 2013, suggesting a process of convergence towards the frontrunner countries. Life expectancy for men, in 2013, was lowest in the Central and Eastern European countries of Bulgaria,

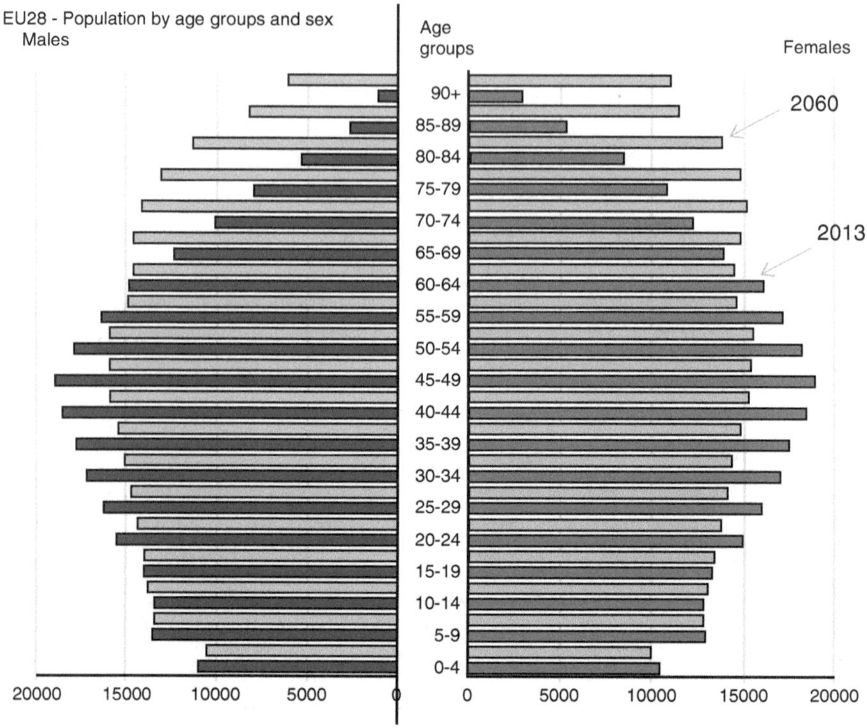

Fig. 1.2 Population by age groups and sex, EU28. (Source: Eurostat 2018)

Estonia, Hungary, Latvia, Lithuania and Romania, ranging between 69 and 72 years. Between 2013 and 2060 life expectancy at birth is projected to rise by more than ten years. For women the largest forecast gains in life expectancy at birth, of eight years or more, occur in Bulgaria, Hungary, Latvia, Lithuania, Romania and Slovakia. In all of them, female life expectancy was below 80 years in 2013 (EC 2014, p. 11). While net migration is highly important in human and policy terms and is projected to aggregate to 55 million people by 2060 (around 10 per cent of the EU population), it is not driving the ageing of the population. This does not mean that net migration does not have significant implications for ageing policies, it does and especially so in those countries with the highest annual inflows: Germany, Italy, Spain and the UK (see Chap. 9). Countries in Eastern Europe (and Central Asia) have been affected also by outward migration which has important personal and policy implications, such as the loss of potential family carers and their social insurance and fiscal contributions. Outward migration also reduces the size of the succeeding generation.

These demographic trends, particularly the continuing ageing of the population, have major policy implications. As those are the subject matter of the rest of this book they will not be elaborated here, but it is clear that they have an impact on a huge swathe of social and economic life: from the labour market to LTC, from economic growth to the family, from technology to health. A central message of this book is that these critical implications can be managed successfully if a strategic approach is taken.

The Active Ageing Index

As has been made clear already active ageing was the key organising concept linking the various strands of MOPACT, and is therefore central to this book. A major task of the core work package (1) was to utilise the AAI as a reference point from which to understand the current status of active ageing outcomes in different Member States of the EU and then to identify specific strategies that policymakers in the different countries might adopt if they want to progress towards the best performing ones. That work is reported in Chap. 2. As a prelude the AAI is introduced here. It was developed for the European Commission and United Nations Economic Commission for Europe during the 2012 European Year of Active Ageing and Solidarity Between Generations (Zaidi 2015). The AAI has been applied in both EU and non-EU countries (Zaidi et al. 2018), although it must be acknowledged to some extent as a work in progress. In particular, the AAI does not yet reflect the life course perspective inherent in the concept of active ageing but, rather, is a measure of policies that support the potential for active and healthy ageing among *older people*. This is valuable in itself and, to some extent, necessary life course policies can be imputed from the outcomes in later life shown by the AAI. Nonetheless, there is an aspiration to transform it into a genuinely life course orientated index (see Chap. 2).

For now we can use the AAI to provide a snapshot of the present state of a large portion of the ageing policy landscape and the variations between the Member States. The AAI comprises 22 individual indicators grouped into four domains: employment, social participation, independent living and capacity for active ageing (Fig. 1.3).

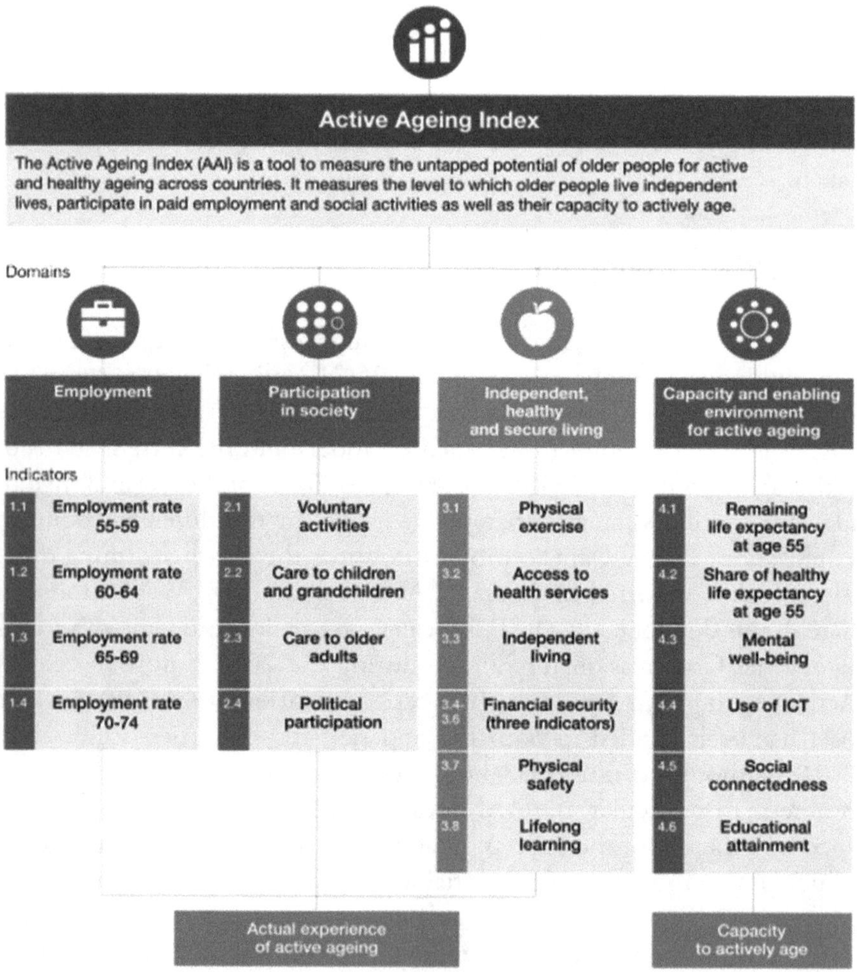

Fig. 1.3 Active Ageing Index—domains and indicators. (Source: Zaidi 2015, p. 3)

The indicators are aggregated to form a single index to show the degree to which a healthy and active life in old age has been realised in different countries. Figure 1.4 shows the average scores for each of the EU28, with the darker shading reflecting higher index values and the lighter shading lower values.

It is clear from the map that Sweden tops the ranking, followed closely by Denmark, Finland, Ireland, the Netherlands and the UK. Four Southern European countries (Italy, Malta, Portugal and Spain) comprise the middle-ranked countries, along with most of the other Western European states. Greece and most of the Central and Eastern European countries are found at the bottom of the ranking. As is explained in

Fig. 1.4 Overall AAI scores for EU28, 2014. (Source: Extended version of the chart included in Zaidi 2015, p. 4)

greater detail in Chap. 2, the high position of the Nordic and Western European countries is primarily due to their policy success in sustaining employment among older workers and in providing income security and access to health care for older people, although, again as explained in Chap. 2, even in these countries there is room for improvement.

Introducing the Book

It will be clear by now that MOPACT was a highly complex multi-dimensional and multi-disciplinary project that has produced a huge array of outputs. It is impossible to encompass that massive body of work in one volume and so the chapters that follow concentrate on distilling the main findings. Following a standard format, each of the chapters provides essential context to the specific aspect of ageing they are addressing and highlights the key steps necessary to ensure the realisation of a transformed future for ageing. Here is a summary of their content.

Chapter 2 does not follow the standard formula of the rest of the book because it is essentially an extension of this one and relies on the contextualisation provided here. Walker and Zaidi distil the main findings from the separate MOPACT work packages and, therefore, provide an overview of the book. They also report the research they carried out on strategies of active ageing in Europe. As indicated above, one of the main objectives of MOPACT was to create an evidence base for developing strategies to promote active ageing as an important way to contribute to the goals of HORIZON 2020 and the EIPAHA. While the 150 exemplars of social innovations in this field are highly relevant to this endeavour, our work on strategies took as its starting point the AAI, as a summary measure to monitor progress across EU countries with respect to active and healthy ageing among older people. Work on the AAI within MOPACT has led to improvement in the fourth domain, which concerns the enabling environment; the identification of indicators to enable a life course perspective within the index; and examination of the unequal experience of active ageing by operationalising the index at the individual level.

Key findings that emerged from the AAI analysis include the fact that active ageing is rising across Europe. However, this positive picture is qualified by the fact that, as in other aspect of ageing policies (see above and Chaps. 5 and 9) the Central and Eastern European countries have weaker institutional capacity and therefore progress lags behind their Western counterparts. To exemplify these variable stages of development of ageing policies and infrastructures across Europe, the chapter examines three clusters of EU countries derived from the above AAI analysis—low, middle and high scoring—and the specific strategies that are required within each of them. For example, low-scoring countries, such as Bulgaria, Croatia, Greece, Hungary, Latvia, Lithuania, Malta, Poland, Portugal, Romania, Slovakia, Slovenia and Spain, would need to take action in the domains of employment opportunities for older workers and social and political participation if they want to improve their active ageing profile. For the middle-scoring countries, such as Germany, France and Italy, it is employment opportunities for older workers that also require most attention. Finally, for the high-scoring countries, including Ireland, the Netherlands, Sweden and the UK, the main areas for improvement are in regard to incomes (especially the UK) and health and social care in Denmark, the Netherlands and Sweden. In their conclusion, Walker and Zaidi argue that it is essential to share good practices across Member States if the key goal of HORIZON 2020 is to be met.

In Chap. 3, Valkonen and Barslund confront some of the enduring and deeply entrenched myths concerning the economic impact of population ageing. In particular, they examine the uncertainties related to the ongoing trends in demography and their implications for fiscal sustainability. A key variable in this analysis is longer lifetimes. The chapter opens with a brief overview of demographic trends which extend the outline earlier in this chapter. With respect to the main economic effects of population ageing, these include lower mortality rates in childhood and later life and their implications for health and social services, and the reduced growth in the consumption of goods and in domestic production. These effects interact with public finances. For example, in public pensions the higher ratio of retirees to workers causes pressures to raise taxes and pension contribution rates even though the needs for child care and education are reduced. The further impact of sustained low growth and close to zero interest rates on government bonds—secular stagnation—have enhanced the focus of policymakers' interest on the economic effects of population ageing.

Set against the gloomy prognoses that follow from this kind of standard economic analysis is the fact that in countries that aged early, such as Japan and Germany, a large portion of the baby boomers are already retired. Yet investment rates are not falling and current accounts are in surplus. Also recently the countries that have experienced the most rapid ageing have grown more than others, perhaps due to increased automation. Another countervailing factor is the rise in the employment rate of older workers, as mentioned above (see also Chap. 5).

Valkonen and Barslund apply new research from MOPACT to this critical set of issues. They argue that, first, the traditional old age dependency measure is not helpful because it excludes large numbers of young people, up to the age of 20. By adopting a prospective economic dependency ratio approach (number of inactive to active persons), they show that the challenge of population ageing is less severe than the traditional measures suggest. This is predicated on the assumption that policy changes will follow the changing age structure of the population. In particular that measures to encourage the further extension of working lives are necessary to maintain current economic dependency ratios.

Chapter 3 also highlights a number of other trends that interact with changes in population structure including, most importantly, the increasing numbers of older people living alone and its consequences for housing, insurance, health and LTC services (see Chaps. 8 and 9). Policy needs to adjust rapidly to these demographic shifts if problems are to be avoided. As the authors argue the ageing population needs to be informed of these predicable risks and what they can do to offset them (see also Chap. 4).

With regard to the central question of public policy concerning fiscal sustainability in the era of longevity, the MOPACT research reported in Chap. 3 is clear: if longer working lives accompany increasing life expectancy, there will be sufficient increases in tax revenues to offset the effects of the growth in expenditures that longer life spans cause. This positive outcome merely requires policies that link eligibility ages for pensions to life expectancy (or possibly HLE) and which seek to maximise health improvements. In response to concerns about equity in the application of such policies, because low-skilled workers are the most likely to be disadvantaged, Valkonen and Barslund make the case for earnings-related pension schemes to be complemented by basic pensions, disability pensions and unemployment insurance. Coupled with the preventative measures proposed in Chap. 7, this analysis establishes the basis for sustainable ageing societies.

Chapter 4 considers ways in which defined contribution (DC) pensions can be made both more intelligible and more effective for contributors and society. As Nijman and Soest note, DC pensions are increasing in prevalence across the globe, with the explicit policy intention of altering the balance between PAYG defined benefit (DB) pensions and DC ones. Chapter 3 addressed sustainability in public DB pensions whereas this chapter focuses on the DC pensions. At the heart of this research was the issue of optimum decision-making because, the authors argue, that is the key to the provision of adequate DC pensions. In particular, it has been demonstrated by previous research that people do not necessarily make the decisions that are in their own best interests, especially when it comes to complicated questions with long-term consequences in a context of substantial uncertainty. So, the central aim of this part of the MOPACT project was to seek ways in which optimal decision-making can be stimulated.

The two specific research questions tackled were: how to optimally contribute to drawdown from invested DC pension capital, and how to structure the decision process taking into account heterogeneity in preferences and circumstances? Substantial academic research analyses and papers were generated in answering these two questions. In practical policy terms, a clear message is the need for restrictions on the drawdown strategy, for example, through caps on the Assumed Interest Rate to ensure stable income streams for retirees. Also, in line with other evidence, Nijman and Soest argue that fine tuning the design of pension products to individual characteristics can produce considerable welfare gains. With regard to the goal of improving people's pension decisions, either in the accumulation or decumulation phases, choice architecture is particularly promising. Financial literacy positively influences involvement with pensions and the quality of pension-related decisions. But improving financial literacy through education is not straightforward because such programmes are not always effective. So, regulations are required to prevent the possibility of choosing inferior pension products or making choices that increase the risk of poverty in old age. For example, this means setting mandatory minimum pension contributions and partner pensions, or an upper limit on the share of risky pension investments permitted in the decumulation phase.

Chapter 5, by Naegele and Bauknecht, examines the issue of extending working lives, which is high on the policy agendas of all European countries, and indeed all developed countries. They address two key questions: how can working lives be extended and what socially innovative, effective and sustainable measures can enable older workers to stay longer in employment? In an invaluable contribution to the literature on late working life, the authors highlight the major lessons to be learned from the various attempts made to date to extend working lives. Their research was conducted in 13 countries in the MOPACT consortium and involved literature reviews, secondary data analysis, policy analysis, the collection of more than 150 case studies of 'promising (good) practices' and a large number of individual and focus group interviews.

Considering the measures taken for extending working lives, Naegele and Bauknecht characterise European countries as either early or late movers: the former being those that have already implemented such policies and the latter those that are just in the process of doing so, and these are largely the Central and Eastern European countries. First of all, they distinguish the policy measures at a national level, such as raising legal retirement ages and closing or making more difficult early labour market exit pathways. They make the important point that not all groups of workers are equally affected by labour market barriers and adaptations to retirement policy. High-skilled employees in favourable working conditions with high salaries have little difficulty remaining in employment, whereas the less skilled and precariously employed face multiple barriers. A strategy of 'work for those who can, social security for those who can't' is crucial, at least in the short term. As well as actions with respect to pension ages and early exit the chapter considers partial pension schemes, working beyond legal retirement age ('silver work'), the promotion of self-employment, employment protection and anti-age discrimination legislation, and emphasises the key lessons from current practice.

The second half of the chapter moves from the national to the meso level and, specifically, the organisational level. Here the authors report their MOPACT research which identified social innovations and good practices undertaken by employers. These range from health promotion, to work-life balance policies, to career planning and the reconciliation of paid work and care. They note that successful innovations are strangely dependent on sector and organisational size. The main drivers of such measures are external circumstances (e.g. lack of skilled labour); funding, especially EU funding; legislation; and collective agreements between trade unions and employers.

After examining both national and meso-level policies and practices, Naegele and Bauknecht conclude that the former can only provide incentives for working longer; it is the latter that are crucial to this endeavour. The chapter then draws to a close with an examination of lifelong learning and underlines its critical part in improving the workability and employability of an ageing workforce. The overall conclusion, based on the evidence, is very optimistic about the prospect for promoting active and healthy ageing in employment.

In Chap. 6, Deeg, Boert and Galenkamp examine healthy life years and social engagement. Their chapter looks at trends in HLE, the correlates of social engagement among older people with and without health limitations, fostering social engagement at the meso and macro levels and the impact of social engagement on HLE. The chapter shows that, between 2007 and 2015 in the EU27, life expectancy (LE) at age 50 increased significantly for both men and women, but LE without activity limitation remained stable. Thus, in 2015, at age 50, on average, men could expect 12.2 healthy life years and women 17.2 years.

Looking at social participation the research found considerable differences in the levels of participation between those with and without multi-morbidity, in favour of the group in better health. There were no major differences, however, in the determinants of participation between those with multi-morbidity and those without—the main factors were socio-economic status, network size, activity in other forms of participation and physical and psychological health status.

A key finding of the research on fostering social engagement at the meso level was the need for improvement in older people's information and communication technology (ICT) skills so that projects aimed at raising participation levels could more easily communicate with potential participants. At the macro level, the researchers undertook case studies in Austria, Italy, Poland and the Netherlands. In all four countries, the policies that most explicitly address the social participation of older people are focused on the labour market. To a lesser extent, there are also policies aimed at raising social participation in the four countries in the domains of caregiving and volunteering but, as the authors point out, encouraging active participation in the labour market may actually interfere with engagement in these other areas. There was also evidence of participation in lifelong learning, via the Third Age Universities in Italy, Poland and the Netherlands. Large areas of potential participation, however, are untouched by policy.

The overall message from this MOPACT research by Deeg and her colleagues is a pessimistic one: the expectation that social participation would be an important factor in improving health, especially for people with multi-morbidity, was not supported. This does not mean, of course, that social participation is not beneficial in other ways for older people and society. Also, there is a clear need for further research on the variety of factors that have the potential to increase HLE—a theme taken up in the next chapter.

Chapter 7 represents the contribution of biologists to the MOPACT project. Bringing them on-board for this multi-disciplinary project was considered essential from the start of thinking about it, but, as noted above, a convincing case had to be made. This chapter demonstrates profoundly the validity of that judgement. In a nutshell, it is impossible to achieve the EU goal of raising HLE without the cutting-edge contribution of biologists, working in partnerships with other disciplines.

Giefing-Kröll and Grubeck-Loebenstein provide a superb summary of the current state-of-the-art with respect to improving health in later life and the specific contribution of their and their colleagues' MOPACT research. They identify the key challenges and especially the failure of HLE to keep pace with LE in Europe. They argue rightly that a life course perspective on healthy ageing is essential to ensure that interventions are made early enough to prevent physical degeneration and the decline of cognitive function. They then apply a novel stages of life approach to illustrate the sorts of interventions required to preserve health and to prevent or ameliorate disease, starting with foetal life and proceeding to old age. A key message of this account is that the acceleration of physical decline caused by external (to the body) factors is generally reversible at any age. Another is that only 20–25 per cent of HLE is predetermined by genes: lifestyle and environment (socio-economic, built and natural) play the major roles. This leads them logically, and in line with the overall MOPACT theme, towards *prevention*. This includes lifestyle changes in, for example, diet and physical exercise, pharmaceutical interventions and the much neglected contribution of vaccination.

Giefing-Kröll and Grubeck-Loebenstein conclude their analysis by emphasising the critical importance of knowledge translation into policy and practice, which is a current major barrier to improvements in HLE, and the need for continuing research in this field.

Chapter 8 covers a broad range of important issues facing European ageing societies: new technologies, housing and transport, all of which are critical to both the ability of older people to live independently and their quality of life. Merkel and his colleagues identify the three key challenges in this broad field: ICTs, age-friendly housing and mobility for all, and provide a state-of-the-art assessment of research knowledge on each of them which draw on case studies of good practices in five countries covered by their research: Finland, Germany, Poland, Spain and the UK. Only a few of their extensive good practice examples could be represented in this chapter, but the rest of them are available on the MOPACT website.

Concerning ICTs, the chapter highlights the significant deployment gap that exists throughout Europe: the proliferation of ICT devices but lack of usage among older people. This is holding back the full development of the silver economy. The barriers such as usability and lack of both awareness and personalisation are mapped not with regard to telehealth and telecare, with guidelines for overcoming them, such as fostering participative design and defining common standards, together with good practice examples. In their conclusion, Merkel and colleagues emphasise the important role of policymakers in Europe, for example, in improving coherence, eliminating obstacles and reducing regulatory complexity and rigidity.

In Chap. 9, Schulmann, Reichert and Leichsenring provide a comprehensive analysis of LTC in Europe, with particular reference to eight countries: Austria, Estonia, Finland, Germany, Hungary, Italy, Portugal and Romania. The additional novel feature of this chapter is the focus on social innovation and active ageing as they pertain to LTC. As the authors point out, both concepts have not been applied hitherto the LTC field.

As a starting point, Chap. 9 identifies the key challenges facing LTC in Europe. On the one hand, all European countries are experiencing the same demographic developments in the extension of longevity coupled with the decline in the birth rate, albeit to varying degrees. But, on the other hand, they do so from highly diverse care regimes. Using the variables demand for care, provision of informal care, provision of formal care and acknowledgement of LTC as a social risk, they distinguish four different care regimes: Nordic (e.g. Finland), standard care mix (e.g. Germany), family based (e.g. Italy) and the transition countries of Central and Eastern Europe (e.g. Hungary). The latter countries in general did not earmark LTC as a priority and, therefore, in terms of the development of LTC provision, those countries are labelled 'late movers' or 'laggards' (see also Chap. 5).

Then the concepts of social innovation and active ageing are outlined and discussed. As pointed out above, 'social innovation' is a new concept and, although much employed in political rhetoric, lacks clarity. So, in its application to LTC, this chapter provides the necessary clarity concerning the criteria that LTC institutions need to fulfil to be classified as social innovations. With regard to active ageing, again as noted already, there is a widespread tendency to consider it as a purely labour market-related phenomenon when, on the contrary, it is much more broadly based (see Chap. 11). This includes the world of LTC which, of course, is chiefly oriented towards dependency and frailty. To draw out the potential for social innovation and active ageing in LTC, Schulmann and her colleagues provide an excellent series of case study vignettes arranged according to care regimes.

The main portion of Chap. 9 is devoted to reporting the findings of the authors' field research, conducted in the eight countries in the MOPACT work package on LTC. This starts with an outline of the drivers behind and barriers facing social innovations in LTC, and then moves on to discuss the possibilities for innovation and realistically implementable scenarios for LTC. Those areas with the greatest potential and need for social innovation are the establishment and expansion of LTC as a system, shifting LTC towards community-based care, facilitating integration and coordination of LTC services, and creating and improving employment in LTC.

In conclusion, Schulman, Reichert and Leichsenring emphasise the stark divide among European countries in terms of social innovation in this field. The Nordic and standard care mix countries are far ahead of the others in terms of innovative practices. So, what are considered as innovations in the transition countries are not classed as such in the more advanced LTC systems. This means that there is a need for much greater sharing of good practice and the pitfalls involved in service delivery, as well as a stronger evidence base for social innovation.

Chapter 10 reports the large block of research conducted by MOPACT on the issue of the political inclusion and participation of older people in Europe. In the first half of the chapter, Moreira and Barslund set out a comprehensive framework of analysis which reflects the fact that the political participation of older people is a function of three sets of mechanisms that operate at the individual, symbolic and institutional levels. Central to this framework is one of the core assumptions of MOPACT that the different dimensions of ageing, including political participation, must be analysed from a life course perspective.

Building on the analytical framework, the second half of the chapter distils the key findings from the MOPACT research, three of which are highlighted. First, the transition to retirement is associated with a decrease in the level of civic engagement. This does not represent withdrawal from public life, however, because other non-employment-related forms of participation appear to remain fairly stable. This analysis also suggests that the influence of cohort membership on political participation may have been over-estimated previously. Second, perceived violations of the consumption, succession and identity norms appear to strengthen negative evaluations concerning the competence of older people and their contribution to the economy. In this process, the role of print media in the reproduction of prescriptive norms about ageing and older people is highlighted, but this influence is less than expected. Third, the involvement of older people in policymaking takes place mainly at the decision-making level, and primarily on a consultative basis. In contrast, their involvement in decisions concerning the implementation of services is rare. In reporting these key findings, Moreira and Barslund pinpoint some priorities for further research.

Conclusion

One of the reasons why some commentators are able, with impunity, to make outrageous claims concerning the damaging impact of population ageing is the absence of accessible robust scientific information. That is why this book is important: it is based on one of the largest and most comprehensive analyses of the policy implications of population ageing ever undertaken and, on that basis, provides ample evidence to undermine the scare mongers and their demography of despair. In particular, the subsequent analyses show that:

- The economic implications of ageing can be managed (Chaps. 3, 4 and 5).
- The decline in HLE can be reversed (Chaps. 6 and 7).
- There is great potential in the silver economy waiting to be unleashed (Chap. 8).
- The potential of LTC, in both employment and well-being, can be realised (Chap. 9).
- The citizenship of the increasingly older population can be enhanced (Chap. 10).

In short, active ageing for all can be a realistic future in Europe.

References

EC. (1999). *Towards a Europe for All Ages*, COM221 (Final). Brussels: European Commission.

EC. (2004). *The 2015 Ageing Report*. Brussels: European Commission.

European Commission. (2013). *Guide to Social Innovation*. Brussels: DG Regional and Urban Policy.

EC. (2014). *Demographic Report*. European Commission: Brussels.

Eurostat. (2018). *Demographic Statistics*. Brussels: Eurostat.

Futurage. (2011). *A Road Map for European Ageing Research*, Sheffield. http://www.futurage.group.shef.ac.uk/home.html

Lawton, M. (1980). *Environment and Aging*. Monterey: Brooks Cole.

Murray, R., Caulier-Grice, J., & Mulgan, G. (2010). *The Open Book of Social Innovation*. London: Young Foundation.

Naegele, G., & Bauknecht, J. (2014). *Conceptual Framework Task 1 of WP3*, MOPACT Project. Available at http://mopact.group.shef.ac.uk/about/

Neary, D., et al. (2015). *A System to Evaluate Social Innovations and to Assess Their Potential Impact on Healthy Life Years*. Sheffield: Department of Sociological Studies.

Oeppen, J., & Vaupel, J. (2002). Broken Limits to Life Expectancy. *Science, 296*, 1029–1031.

Pol, E., & Ville, S. (2009). Social Innovation: Buzz Word or Enduring Term? *Journal of Socio-Economics, 38*(6), 878–885.

Sidrenko, A., & Walker, A. (2004). The Madrid International Plan of Action on Ageing: From Conception to Implementation. *Ageing and Society, 24*(2), 147–165.

UN. (2002). *Report of the Second World Assembly on Ageing*. New York: United Nations.

Wahl, H.-W., & Iwarsson, S. (2007). Person-Environment Relations in Old Age. In R. Fernandez-Ballesteros (Ed.), *Geropsychology: European Perspectives for an Ageing World* (pp. 49–66). Göttingen: Hogrefer.

Walker, A. (2009). The Emergence and Application of Active Ageing in Europe. *Journal of Aging and Social Policy, 21*, 75–93.

Walker, A., & Foster, L. (2013). Gender and Active Ageing in Europe. *European Journal of Ageing, 10*(1), 3–10.

Walker, A., & Maltby, T. (1997). *Ageing Europe*. Buckingham: Open University Press.

WHO. (2002). *Active Ageing: A Policy Framework*. Geneva: World Health Organisation.

Zaidi, A. (2015, April). Active Ageing Index: *A Legacy of the European Year 2012 for Active Ageing and Solidarity Between Generations*. Policy Brief, Vienna, European Centre.

Zaidi, A., & Stanton, D. (2015). *Active Ageing Index: 2014: Analytical Report*. Brussels: UNECE and European Commission Directorate General for Employment, Social Affairs and Inclusion.

Zaidi, A., Gasior, K., Marin, B., Rodrigues, R., Schmidt, A., & Zolyomi, E. (2017). Measuring Active and Healthy Ageing in Europe. *Journal of European Social Policy, 27*(2), 138–157.

Zaidi, A., Harper, S., Howse, K., Lamura, G., & Perek-Białas, J. (2018). *Building Evidence for Active Ageing Policies: Active Ageing Index and Its Potential.* Basingstoke: Palgrave Macmillan.

2

Strategies of Active Ageing in Europe

Alan Walker and Asghar Zaidi

Introduction

As explained in the previous chapter, one of the core endeavours of the Mobilising the Potential of Active Ageing in Europe (MOPACT) project was to generate a strong evidence base for developing strategies to promote active ageing and thus help achieve the goals of Horizon 2020 and the European Innovation Partnership on Active and Healthy Ageing (EIP-AHA) policy frameworks. In this process, the broader societal goal of empowering older people and their communities was also pursued.

A. Walker (✉)
Department of Sociological Studies, University of Sheffield, Sheffield, UK
e-mail: a.c.walker@sheffield.ac.uk; https://www.sheffield.ac.uk/socstudies/staff/staff-profiles/walker

A. Zaidi
Centre for Analysis of Social Exclusion (CASE),
London School of Economics and Political Science, London, UK
e-mail: A.Zaidi@lse.ac.uk

A. Walker (ed.), *The Future of Ageing in Europe*,
https://doi.org/10.1007/978-981-13-1417-9_2

MOPACT promoted a consultative method by engaging key stakeholders and making research relevant and effective.

The active ageing strategies point to the steps necessary to close the so-called structural lag in policies and institutions which have lagged behind longevity so that the structural changes required to maximise its positive impact have been slow to materialise (Walker 2018). The steps identified will need to factor in the differential capacity for active ageing on the basis of the cultural norms and institutional diversity across the European Union (EU) countries, so as to point out how active ageing can be promoted within varying national and regional contexts (Walker and Zaidi 2017).

The development of active ageing strategies assembled information from various streams of research within MOPACT. The critical research questions addressed include what policy-relevant knowledge is emerging from the literature review and empirical work undertaken across the research packages in MOPACT? What specific policy priorities can be identified for Europe as a whole? What challenges, and in what forms, need to be addressed in the implementation of these strategies in the future? What distinctive policy strategies emerge for individual countries from the analysis of the main clusters of EU countries?

As shown in Chap. 1, the starting point for the research was provided by the work undertaken in the EC/UNECE's Active Ageing Index (AAI) project (Zaidi and Stanton 2015; Zaidi et al. 2018). The AAI project, which started as one of the activities of the European Year for Active Ageing and Solidarity between Generations (Decision 940/2011/EU), offered evidence on the extent of active ageing experienced by older populations across EU countries which was further strengthened in MOPACT.

The active ageing definition followed emphasised particularly the importance of enabling environments and opportunities to engage in wider forms of activities and relationships:

> ...the situation where people are able to live healthy, independent and secure lives as they age and thus continue to participate in the formal labour market as well as engage in other unpaid productive activities (such as volunteering and care provision to family members). (Zaidi et al. 2013, p. 6)

This definition was derived from the World Health Organisation (WHO) one quoted on page 6 in Chapter 1. It captures the essence of the WHO definition in being concerned with the ageing process and enabling environment but we acknowledge the point made in the previous chapter that its emphasis is primarily on the latter part of the life course. Our work within MOPACT has started to refocus the AAI on the whole of the life course, a project that is ongoing.

The AAI is a summary measure that monitors progress across EU countries with respect to active and healthy ageing of its older population (in most instances referring to the age group of 55+). It provides an assessment of the untapped potential among older people using 22 indicators that are grouped in four domains: employment, social participation, independent living, and capacity for active ageing. As the ageing experiences of men and women are expected to be different, the AAI also provides a breakdown by gender which points to inequities in experiences of active and healthy ageing within EU countries.

MOPACT generated evidence on what revisions are necessary in this choice of indicators and domains and on what grounds. The following improvements had been recommended for the AAI framework from the research undertaken in MOPACT.

1. Updating and extension of the fourth domain of the AAI framework, namely, enabling environment for active ageing, where additional indicators capturing the capacity and enabling environment for active ageing have been considered.
2. Identifying and operationalising active ageing using the life course perspective, with a focus on early life experiences and resources and linking them to active ageing outcomes at older ages.
3. Examining unequal experiences of active ageing by operationalising AAI at the individual level.

Further research was also undertaken in MOPACT about whether the findings of the AAI reflect the actual situation and what aspects require a greater scrutiny. Also, how variations across countries are associated with different social policy contexts, perceptions, cultures, and institutions, and what further analysis across subgroups (on the basis of age, gender, disability, employment status, marital status, living arrangements, and educational attainment) is required to understand better the differences across countries.

Policy-Relevant Knowledge Synthesised from MOPACT

The research undertaken in each of the work packages draws out key conclusions about what challenges lie ahead, what policy reforms are required, and at what level. The suitability of these policy actions is then assessed bearing in mind the different cultural norms and institutional settings across the EU countries (in some cases with the help of pairwise comparisons of EU countries) (see, e.g. Kureková and Zólyomi (2016) for a comparison between Czechia and Poland; and Scoppetta (2016) for Austria and Germany). To determine what policies and programmes are required, it is also essential to determine the priorities, the aspirations, and relevance for each group of countries.

Challenges Ahead

The challenges presented by population ageing in European countries are numerous, both economic and non-economic. The overall message is that these challenges are not insurmountable but require new thinking, especially by policymakers, about ageing and what it means (Walker 2014; Walker and Zaidi 2017). The economic challenges include the future sustainability of public welfare systems, and they are covered in detail in Chaps. 3 and 5 of this book. The non-economic challenges to overcome are summarised below.

Old and Living Alone

One of the future challenges facing the European societies is a fast-rising proportion of people aged 80+ who live alone (Keilman 2015). It is therefore important to look into the special needs and aspirations of this specific group of the population in improving their experiences of active and healthy ageing. Individuals should be informed about the increased economic and social risks of living alone and for longer than expected, and societies should prepare for likely multiplication of the numbers of older citizens in one-person households. Strategies should be devised to engage older persons in social and cultural activities, aimed at their empowerment and greater participation in the communities in which they live.

Healthy Life Expectancy

Another challenge is to (continue to) introduce measures that increase healthy life expectancy alongside the rising life expectancy (Luijben et al. 2014). For this purpose, a greater emphasis needs to be placed on upstream, early-in-the-life-course interventions that improve healthy ageing during older ages (WHO 2012), and greater focus on the countries with the largest differences between life expectancy and healthy life expectancy (Slovakia, Portugal, Germany, Slovenia, Estonia, Spain, and France), as noted in Chap. 7.

Many molecular and cellular changes which take place during ageing are well understood: only 20–25% of healthy life expectancy is predetermined by genes, and it is therefore the lifestyle and environment that are critical determining factors for active and healthy ageing, as discussed in Chap. 7.

To improve and define new early intervention strategies, it is vital to develop excellent translation practices. For example, novel strategies (e.g. dietary interventions, novel drugs, stem cells, personalised medicine) need successful translation from the understanding of molecular mechanism to animal models to clinic (Lepperdinger 2016) and subsequently changes in the living habits of humans.

Long-Term Care

Long-term care has been increasingly acknowledged as a social risk and it is emerging as an institution in its own right in most countries. In this respect, important distinctions are critical between health and social care and between formal and informal care.

Long-term care policy is found to be merely a process of 'muddling-through' in many countries—with clear strategies and objectives missing and policymakers paying only lip-service to the implementable ideas of 'ageing in place'. A coherent policy design and the institutional setup for the provision of long-term care are therefore needed in many EU countries (Chap. 9 and Leichsenring 2016).

Active Citizenship and Political Participation

There is a need to significantly increase the opportunities for seniors to be involved in the decisions about the delivery of services. Currently, the evidence suggests that the direct involvement of seniors in policymaking is an exception rather than rule. Where it happens, it is mostly done through the creation of consultative bodies (e.g. Seniors' Councils or Seniors' Forums) at the local level (Chap. 10 and Moreira 2016).

Tapping the Silver Economy

Measures tapping the unfulfilled potential of information and communication technology (ICT) use will be effective in promoting active and healthy ageing. An increasingly greater use of the ICT by the current and future generations of older people offers a great scope for innovations and improving their cost-effectiveness in the future. The silver economy potential has been largely untapped in many countries, due principally to low user involvement and administrative constraints (Chap. 8 and Hilbert 2016).

Unleashing Social Innovation

As indicated in Chap. 1, social innovation has the promise to assist the active ageing transformation. Already there are a wide range of social innovation projects. In recent work, we have together with collaborators documented and assessed a wide range of social innovations so that policymakers and social innovators can be informed and inspired about ideas, concepts, and projects that are operating elsewhere in the world and could be developed in other places (Walker et al. 2017). The social innovations were summarised and assessed using a balanced scorecard containing four domains, namely: impact, sustainability, implementation, and transferability—on a simple 1–5 scale with a minimum standard of 12 for inclusion in the list. Examples span social innovations across the four domains of the AAI: employment; participation in society; independent, healthy and secure living; and capacity and enabling environment for active ageing (MOPACT 2016; Neary et al. 2015).

Key Policy-Relevant Findings from MOPACT

Active Ageing Rising Across EU Member States

The evidence drawn from the analysis of the AAI offers a clear sign of a change that there is now a greater emphasis on active ageing outcomes across Europe. Still, there remains a significant scope for further improvement, even in countries which had a greater success in promoting active ageing so far. For most countries, the changes for both men and women showed improvement although with a significant gender gap in almost all countries.

Moreover, all other things being equal, the individual-level experiences of active ageing have become more equal across subgroups in the recent past (for the nine countries where relevant temporal data is available) during the period from 2004 to 2013 (Barslund et al. 2017). Further analysis shows that a higher experience of active ageing is correlated with a higher quality of life of older people. Likewise, the relationship between active ageing and per capita wealth is also significantly positive. Thus, active ageing does not imply a worsening of older people's quality of life and it brings real benefits to the economy (Walker and Zaidi 2016).

Greater Challenges for Central European Countries

Active ageing in central European countries is typically low, reflective of weaker institutional capacity and enabling environment for active ageing in these countries. In particular, social participation of the older population is markedly low.

Employment among older persons is relatively higher which reflects inadequate levels of pension income constraining people to remain longer in employment. When the problem of pension income adequacy is addressed, the higher levels of employment (especially among people over retirement age) will not be sustainable without further supportive policy initiatives improving the work environment and pension income incentives for longer working careers.

The central European countries with low active ageing outcomes also show higher inequality across subgroups. This finding shows that helping the most vulnerable in EU Member States with low active ageing will go a long way towards supporting more equality across Member States.

Greater Risks of Social Exclusion in Older Age

Overall, life expectancy at age 65 in EU countries increased, but healthy life expectancy decreased marginally. Similar trends were observed at age 50 and at age 85. At the same time, evidence point to a larger variety between the countries in healthy life expectancy and trends in healthy life expectancy. These findings imply that greater time will be spent in the future in a health status that requires specialised care and support.

One other source of greater risks of social exclusion in the future is a change in the age structure of the European population, with a rising proportion of one-person families. The number of people 80+ living alone will grow by at least twofold in 25 years (in the five countries studied). Individuals and societies should be prepared for the increased economic and social risks of living alone in later life.

Ageing Is Not Inevitably Linked to Chronic Diseases

MOPACT evidence points to the fact that only 20–25% of healthy life expectancy is predetermined by genes. In fact, age-associated molecular and cellular damage can be prevented or greatly delayed by lifestyle interventions (e.g. dietary manipulations) and by enabling environments (e.g. knowledge of nutrition, facilities for exercise, timely vaccination). Social investments are required in earlier phases of life to maximise the healthy part of longevity which is subsequently responsible for a variety of activities in later life.

EU Institutions Playing an Important Role but More Is Required

The funding for MOPACT has generated strong scientific evidence to inform public policy priorities for active ageing in Europe. Strong political commitments from national and EU-level policymaking institutions are now required for social policy reforms to realise active ageing at all ages. MOPACT provides a new vision of ageing in which longevity is being considered an asset for social and economic development. Similar multidisciplinary projects on active ageing require further funding and translational research towards the formulation of strategies.

Five Prerequisites for Realising Active Ageing in Europe

- First, the optimistic paradigm of ageing needs to be promoted, viewing older people as agents of change, and with social and economic rights.
- Second, a life course perspective needs to be adopted, putting a strong emphasis on active ageing at all ages.
- Third, heterogeneity among older persons needs to be reflected in policies. In the process, the active ageing strategies should be made relevant for all and not reduced to elites who are better educated and enjoy good health in later life.
- Fourth, the multidimensionality of active ageing must capture all its aspects of economic and social activities. Active ageing should not be reduced to the idea of longer working careers but also social engagement, independent living and security.
- Fifth, in devising active ageing strategies, it is essential to take account of diversity in the institutional capacities and enabling environments for active ageing across EU Member States.

Major Policy Priorities and Instruments

MOPACT points to the various top-level policy goals in European countries. These goals are outlined below along with a description of policy reforms necessary.

Goal 1: Ensuring Economic and Social Sustainability

The key to economic sustainability is linking retirement ages to life expectancy. This strategy, when pursued by keeping account of life expectancy differences across different socio-economic classes (Zaidi and Whitehouse 2009), will also contribute to reducing inequalities in pension income outcomes within a country.

To extend working lives, it is essential to coordinate public policy with workplace human resource policies. Furthermore, interventions need to be tailored to match the state of improvement in individual health and other aspects of readiness in each country and organisation.

To stimulate private savings, the mere provision of information and financial literacy is not enough, but there is a need to provide specific guidance of groups at risk of low savings during their working lives.

All policy reforms towards financial sustainability must keep in sight the goal of maintaining pension income adequacy. Keeping the poverty levels of older people in check will contribute strongly to the sustainability of social welfare systems.

Goal 2: Promoting Health and Well-Being in Later Life

Urgent actions are required in this respect in countries where the gap between life expectancy and healthy life expectancy is largest: Slovakia, Portugal, Germany, Slovenia, Estonia, Spain, and France.

The classic public health interventions, such as promoting nutritional diets, exercise, and vaccination, are required in all EU Member States to promote health and well-being outcomes in later life.

Social investment type early life interventions are required in increasing healthy life expectancy to keep pace with rising life expectancy and reduce disability and dependency in later life.

Goal 3: Provision of Adequate Person-Centred Long-Term Care

Interventions are required on several fronts in improving long-term care services.

- Expanding existing services where the provision is inadequate;
- Greater coordination and integration of health and social care services;
- Innovative design and implementation of community care; and
- Employment and professional training of carers.

In this respect, there is a clear divide between Member States in their readiness for and implementation of much needed social innovations in long-term care. Therefore, there is a greater scope for knowledge sharing between the frontrunner countries and other countries in Europe and the rest of the world.

Goal 4: Adopt Modern Technologies for Efficiency and Effectiveness

ICT-based assistive technologies have huge potential for active ageing and the silver economy, but the deployment gap needs to be reduced. The key barrier for an effective deployment is lack of awareness and acceptance among users. Furthermore, public subsidies are required to incentivise the take up of the ICT-based assistive technologies.

Goal 5: Engagement of Older Persons in Social, Cultural, and Civic Activities

Active ageing strategies should pay particular attention to promotion of social engagement and cultural and civic activities in later life. Special emphasis needs to be placed on lifelong learning activities.

Greater participation of older persons requires commitments from and coordination between political authorities and civil society. For instance, there is a need to create local advisory bodies in all Member States to represent the views of seniors.

The heterogeneity of older people (age, education, gender, class, culture) has to be taken into account if participation initiatives are to be successful. For instance, the strategies should place a particular emphasis on older persons living alone.

Strategies for Clusters of EU Countries

Policy priorities for individual countries can also be examined on the basis of the strengths and areas of improvement for each country delineated by the AAI—(Fig. 1.2, page 11). The commonalities for the three groups of EU countries will help further in identifying strategies to promote active ageing in these countries. In particular, we will point to some specific proposals for the New Member States which are identified to be lagging behind.

Low-Scoring Countries

The first group of countries analysed here are those which have scored low in the overall AAI—they are Bulgaria, Greece, Spain, Latvia, Lithuania, Hungary, Malta, Poland, Portugal, Romania, Slovakia, Slovenia, and Croatia. The strengths and areas for improvement for these countries are presented in Table 2.1.

The AAI values within this group show that the Central European New Member States cluster together with Southern European countries, with the AAI values ranging from 28.1 to 31.6. Three countries perform noticeably worse within this group: Greece, Poland, Hungary, and Slovakia. Estonia and Czechia are two exceptional Central European countries as they belong to the middle-scoring countries. Czechia was among the first countries in Central European countries to initiate a nation-wide programme on ageing, the National Programme of Preparation for Ageing, for the Years 2003–2007 (Perek Bialas et al. 2006). It also embarked on a more specific programme to promote active ageing, namely National Action Plan Supporting Positive Ageing, during 2013–2017.

Table 2.1 Strengths and weaknesses identified based on the latest results of the Active Ageing Index (AAI), for *countries with low scores*

AAI rank	AAI value	Country	Strengths	Potential areas for improvement
16	33.7	Portugal	Above average employment rates for those 65+; and higher informal care provision to older adults; excellent social connectedness	Political participation, voluntary activities (especially for men), healthy life years (especially for women), physical exercise and use of ICT (especially for men)
17	32.6	Spain	High life expectancy; many aspects of active ageing close to the EU average, except falling short in employment of older workers	Employment and independent living arrangements
18	31.6	Croatia	Just above average social participation, good at keeping material deprivation low	Employment, poverty risk, lifelong learning and use of ICT
19	31.5	Latvia	Rather high employment, especially among women; and excellent educational attainment	Social participation (especially among men); meeting health and dental care needs, independent living arrangements, material deprivation, physical safety; capacity for active ageing, in particular life expectancy, and mental well-being
20	31.5	Lithuania	Just above average employment (higher for women); above average care to children/grandchildren and older adults	Voluntary activities and political participation; material deprivation, physical safety, lifelong learning
			Good at physical exercise, meeting health and dental care needs, independent living arrangements; high educational attainment among older people	Capacity for active ageing, particularly, life expectancy, mental well-being and social connectedness
21	31.5	Malta	Average social participation, independent living and capacity for active ageing scores, excellent healthy life years and meeting medical needs	Employment, especially among women, and political participation

(continued)

Table 2.1 (continued)

AAI rank	AAI value	Country	Strengths	Potential areas for improvement
22	29.9	Bulgaria	Low life expectancy, but many healthy years in the remaining life expectancy Good educational attainment for the older population	Employment, voluntary activities and political participation; physical exercise, poverty; life expectancy and use of ICT
23	29.8	Slovenia	Excellent educational attainment, met medical care needs and good in physical safety	Employment (for the two younger age groups 55–59 and 60–64, especially women) Also, political participation, physical exercise, mental well-being and use of ICT
24	29.6	Romania	Above average employment, especially at higher ages Good relative median income	Life expectancy, mental well-being, use of ICT and social connectedness; lifelong learning, unmet health care, and physical exercise Voluntary activities, political participation among women
25	28.5	Slovakia	Good care to children and grandchildren; low risk of old age poverty Good educational attainment for the older population	Employment; voluntary activities and political participation Lifelong learning; (healthy) life expectancy
26	28.3	Hungary	High relative income; low risk of old age poverty Also good care to children and grandchildren	Employment (in particular in the age group 60–64), as well as voluntary activities and political participation, lifelong learning, life expectancy and social connectedness
27	28.1	Poland	Good educational attainment for the older population and relative median income of the elderly	Across the board, especially employment, lifelong learning and independent living arrangements
28	27.6	Greece	Relative median income highest, although attributed to low levels of working age incomes Good care to children and grandchildren	Across all four domains, particularly employment, voluntary and political participation, lifelong learning, use of ICT, mental well-being and social connectedness

Source: Zaidi and Stanton (2015), pp. 24–25

The commonalities arising from the Central European countries in this low-scoring cluster can be summarised as:

- Latvia, Lithuania, and Romania have above average employment scores. Since these countries also show a low position in the indicators of median disposable income of people aged 65+ relative to those aged below 65, the high employment rates point to low pension income constraining people to delay their retirement. These countries require further policy initiatives to improve pension income adequacy for their older populations, especially in Bulgaria and Romania.
- The majority of other countries exhibit low employment levels for older workers, particularly the three bottom-ranked countries Poland, Hungary, and Slovakia. Clearly, there is a need to improve employment opportunities for older workers through pension incentives and by introducing age-friendly work environments.
- All the Member States in this cluster have low scores in the social participation domain. This is despite the fact that these countries do very well with respect to the educational status of older persons. This situation reflects the loss of social capital during the past 40 years of communist occupation,[1] but it may also reflect low income levels for the older population as mentioned above. A common standout priority for all these countries is therefore actions to empower older persons with better incomes and age-friendly environments, to raise their social and civic participation.
- There is one change where an almost universal decline confirms a trend across virtually all the New Member States—political participation. This could be due to a widespread disillusion with traditional political parties, which has resulted in a fall in participation in political activities during the period 2008 to 2012.

In Poland, employment among older workers aged 55+ has been one of the lowest in the EU countries. Their 'Solidarity across Generations' programme, which started in 2008 and is planned to run until the end of 2020, shows good early results to address the problems of maintaining people over 50 in employment. It comprises a combination of active labour market measures (e.g. hiring subsidies, job placement, career counselling and training) and also health prevention and early intervention measures.

The results of the AAI have been widely discussed by policymakers in Malta leading to the development of the 'National Strategic Policy for Active Ageing: Malta 2014–2020'. The strategy follows the AAI domains with policy recommendations for employment, social participation, and independent, healthy and secure living. The strategy also presents descriptive analyses of trends and issues that arise in each of these crucial active ageing areas. A distinct unit within the Department for the Elderly and Community Care has been set up to implement this policy.

Middle-Scoring Countries

The middle-scoring countries cover a surprisingly heterogeneous group. This cluster includes the three major economies of Europe: Germany, France, and Italy. Two better performing Central European nations also belong to this cluster: Czechia and Estonia. The other four countries included in this cluster are smaller population ones: Belgium, Luxembourg, Cyprus, and Austria. The weaknesses and areas for improvement with respect to active ageing for each of these countries are reported in Table 2.2.

The following commonalities and policy strategies can be identified for this group.

- In two of the most prosperous EU countries, Germany and Luxembourg, the social care contribution of older persons is relatively weak. This may partly be due to the fact that these countries fare well in their provision of formal social care to older adults. Estonia is quite similar in this respect, as older persons fall short of providing care to other adults in comparison to other countries in this group.
- In comparison to high-scoring countries, many countries in this group need to improve employment among older workers (in particular Austria, France, and Luxembourg). Czechia needs to do better in promoting employment among older women.

Table 2.2 Strengths and weaknesses identified based on the latest results of the Active Ageing Index (AAI), for *countries with middle scores*

AAI rank	AAI value	Country	Strengths	Potential areas for improvement
7	35.8	France	Independent and secure living very high; high life expectancy	Employment among the over-60s and lifelong learning
8	35.7	Luxembourg	Independent living very high, good also in the use of ICT	Employment and care to older adults
9	35.4	Germany	High employment, independent living arrangements and material well-being, along with educational attainment	Social participation, in particular care to older adults and to children and grandchildren
10	34.6	Estonia	High employment, especially among women	Social participation, especially as regards care to older adults and children/grandchildren and political participation. Also, relative median income and capacity for active ageing, particularly share of healthy life years, mental well-being, and social connectedness
11	34.4	Czechia	Excellent educational attainment and low poverty risk	Female employment, life expectancy and physical exercise
12	34.2	Cyprus	High male employment and care to children, grandchildren	Relative median income and capacity for active ageing, in particular social connectedness, use of ICT and mental well-being
13	34.1	Austria	Good in the social participation domain, especially in voluntary activities; independent living, and also meeting medical need and combating poverty; capacity	Employment and care to children and grandchildren

(*continued*)

Table 2.2 (continued)

AAI rank	AAI value	Country	Strengths	Potential areas for improvement
14	34.0	Italy	High in the social participation domain, especially in care of children and grandchildren	Employment, especially among women, and physical exercise
15	33.7	Belgium	Good in the independent living and capacity for active ageing domains, especially with respect to health care needs and material deprivation, in spite of low relative median income	Employment, especially in the age group 60–64, and lifelong learning

Source: Zaidi and Stanton (2015), pp. 24–25

The policy goal of extension of working career requires adjustments in terms of flexibility from employers as well as health promotion measures for older workers. Some good policy practices come from Austria, 'Fit2Work Programme'; from Germany, 'Perspective 50+ Programme'; and from the Czechia, 'Go 50 Go programme' can be replicated in other EU countries. These programmes offer support in keeping people in the job despite health-related problems and in integrating long-term unemployed back into the labour market. For instance, the Austrian Fit2Work programme, introduced in 2011, has been offering free advice and services to support workers and employers seeking to prevent occupational disability and to enable returning to work for those who have experienced long-term absence from work for health reasons.

Spain can be seen as an outlier in this cluster of countries, especially since it is positioned highly (12th) in the AAI domain 'Capacity and enabling environment for active ageing', achieving a position above the average for the EU countries. Their recently formulated active ageing strategy, set out in the document Operational Framework for Elderly People, represents a further step in the evolution of the active ageing policies drawn up and implemented in Spain previously. The strategy can be praised for the active consultation with the older people, through the State Council of Elderly People, the Spanish government's consultative body for policies connected with this social group. Furthermore, many autonomous regional governments have, in accordance with the policy lines established at the national and European levels, devised their own active ageing strategies, for example, The Active Ageing Plan of Biscay makes use of the AAI as the tool to promote specific active ageing strategies and their monitoring (Madrid, IMSERSO 2014).

Czechia's National Action Plan Supporting Positive Ageing (2013–2017) can be mentioned as a good policy example for its focus on securing and protecting the human rights of older people, lifelong learning, employment of older workers, age-friendly environments for seniors, volunteering, healthy ageing and elder care. One of its tasks has been to create a set of indicators that will allow the progress of active ageing policies in the country to be monitored, and the AAI indicators could complement these efforts. Implementation of this programme would have benefited greatly from better financial resources and sufficient government commitment to the cause of active ageing among older population (Koucká and Kučera 2014).

High-Scoring Countries

The high-scoring countries comprise the three Nordic countries (Sweden, the Netherlands, and Finland) and two Anglo-Saxon countries of the EU (UK and Ireland). Their results are included in Table 2.3.

As this cluster comprises the top scoring countries, the urgency for action is not as great as for the two other clusters. There are, however, policy areas that require attention in order to sustain and build on the high level of performance:

- All countries in this cluster perform well with regard to employment except Ireland, where action is clearly needed to raise the level of this domain to match its achievement in the other three. The good employment practices of the rest of the cluster would be a sound basis for Irish policymaking.

Table 2.3 Strengths and weaknesses identified on the basis of the latest results of the Active Ageing Index (AAI), for *countries with high scores*

AAI rank	AAI value	Country	Strengths	Potential areas for improvement
1	44.9	Sweden	Excellent across the board, particularly in employment, voluntary and political participation and the ICT use	Relative median income and unmet healthcare needs
2	40.3	Denmark	Excellent employment rates across the age range, especially beyond the age threshold of 65; political participation among the highest in the EU; low risk of old age poverty; high physical safety and lifelong learning	Relative median income, care to children and grandchildren and care to older adults
3	40.0	Netherlands	Excellent across the domains, in particular male employment. Also, among the top-level countries with respect to voluntary activities; political participation above average. Low risk of old age poverty, high physical safety, and lifelong learning	Female employment, care to children and grandchildren. Relative to the high performing Nordic countries, physical exercise and political participation
4	39.7	UK	Excellent employment rates across the age range; social participation also good across all its aspects (especially for women); also good in independent living arrangements in meeting health care needs and in the use of ICT	Risk of old age poverty and material deprivation
5	39.0	Finland	Excellent across the board, especially in independent living, physical exercise, mental well-being, use of ICT	Male employment and relative median income
6	38.6	Ireland	Highest ranked in the social participation domain, also low unmet health care needs and material deprivation	Employment

Source: Zaidi and Stanton (2015), pp. 24–25

- Despite overall high scores income levels of older people require attention, especially in the UK.
- Health and care needs are areas of under-performance in Sweden, Denmark, and the Netherlands.

Conclusions

This chapter reports the work in MOPACT orientated towards the development of evidence-based policy strategies to promote active ageing and to assist the process of making longevity an asset. Two sets of evidence were adduced to further this major task. On the one hand, we synthesised the mountain of state-of-the-art evidence generated by the other MOPACT work packages, which in turn was subjected to close scrutiny and interrogation by stakeholders—a repeated process that was built into the MOPACT architecture from the start (Fig. 1.1, page 3).

On the other hand, we utilised the EC/UNECE's AAI and undertook analysis of the situation in individual countries to highlight the critical foci for policy action. This work enables us to identify strategies both for individual countries and for their clusters by which their active ageing performance may be raised, not just to improve each country's national ranking on the AAI league table, but also contribute to the overall EU goal of raising healthy life expectancy. Here a key focus was on the 'structural lag' whereby policies on ageing and institutional change fail to keep pace with demographic developments—closing this gap is essential if the goal of Horizon 2020 is to be achieved.

In synthesising policy-relevant knowledge from the whole range of MOPACT work streams, six key policy challenges were highlighted:

- The rising population aged 80+ living alone
- The need to increase healthy life expectancy
- Long-term care systems to improve
- Active citizenship promoted
- Tapping the silver economy
- Maximising social innovation

Within the context of generally rising active ageing, five specific policy priorities were identified:

- Ensuring economic and social sustainability
- Promoting health and well-being in later life
- Provision of adequate person-centred long-term care
- Adoption of ICT-based assistive technologies
- Engagement of older people in social, cultural, and civic activities

Finally, this chapter provided policy strategies geared to the three major clusters of EU countries based on the AAI evidence. The evidence presented will enable the cross-country sharing of good practice policies and mutual learning between EU Member States as they strive to improve their performance in terms of active ageing. The information distilled in this chapter will be of assistance to policymakers throughout the EU in framing recommendations designed to achieve the key goal of Horizon 2020 and the ultimate goal of making longevity an asset for Europe.

Note

1. Many of the former Soviet occupied countries of this region experienced a systematic attempt to destroy the existing institutional arrangements that underpin social capital (see Applebaum 2012). This will affect present day measures designed to support social participation among the elderly and there will be long lags in between initiating policies and seeing social capital rebuilt in these countries.

References

Applebaum, A. (2012). *Iron Curtain: The Crushing of Eastern Europe 1944–56*. London: Penguin Books.

Barslund, M., Werder, M., & Zaidi, A. (2017). Inequality in Active Ageing: Evidence from a New Individual-Level Index for European Countries. *MOPACT, Ageing and Society*, 1–27. https://doi.org/10.1017/S0144686X17001052

Hilbert, J. (2016). *MOPACT Workpage7: Built and Technological Environment.* MOPACT Active Ageing Forum, Vienna.

Madrid, IMSERSO. (2014). Peer Review: Spain. *The Active Ageing Index and Its Extension to the Regional Level.* Madrid.

Keilman, N. (2015). Probablistic Household Forecasts for Five Countries in Europe. *MOPACT-FP7 Project: Delivery 2.5 of Work Package 2 'Economic Consequences of Ageing.*

Koucká, M., & Kučera, T. (2014). Peer Review: Czech Republic. *The Active Ageing Index and Its Extension to the Regional Level.*

Kureková, L. M., & Zólyomi, E. (2016). 'A Country-Pair Comparison of the Czech Republic and Poland', WP1, Contribution to MOPACT Deliverable D1.5 'Scenarios of Active Ageing Document Providing Policy and Practice Guidelines on How to Make Longevity an Asset for Social and Economic Development'.

Leichsenring, K. (2016). *Social Support and Long-Term Care for Older People-Potentials for Social Innovations in Active Ageing.* MOPACT Active Ageing Forum, Vienna.

Luijbn, A. H., Galenkamp, H., & Deeg, D. (2014). *Mobilising the Potential of Active Ageing in Europe-Trends in Healthy Life Expectancy and Health Indicators Among Older People in 27 EU Countries.* MOPACT FP7 Project.

MOPACT. (2016, May 17). *Social Innovations Database.* Retrieved from http://mopact.group.shef.ac.uk

Moreira, A. (2016). *How Can We Promote the Participation of Seniors in Policy-Making Processes?.* MOPACT Active Ageing Forum, Vienna.

Neary, D., Walker, A., & Zaidi, A. (2015). *Social Innovation and Active Ageing in Europe.* MOPACT-FB7 Project WP1.

Perek-Białas, J., Ruzik, A., & Vidovićová, L. (2006). Active Ageing Policies in the Czech Republic and Poland. *International Social Science Journal, 58,* 1468–2451. https://doi.org/10.1111/j.1468-2451.2008.00655.x.

Scopetta, A. (2016). Country-Pair Comparison of Austria and Germany. *WP1, Contribution to MOPACT Deliverable D1.5 'Scenarios of Active Ageing Document Providing Policy and Practice Guidelines on How to Make Longevity an Asset for Social and Economic Development'.* Vienna: European Centre of Social Welfare Policy and Research.

Walker, A. (2014). *The New Science of Ageing.* Bristol: Policy Press.

Walker, A., & Zaidi, A. (2016, May). New Evidence of Active Ageing in Europe. *Intereconomics, 51*(3), 139–144.

Walker, A., & Zaidi, A. (2017). *A New Approach to Ageing in Europe: Making Longevity an Asset.* MOPACT Policy Brief: 10, University of Sheffield.

Walker, A (2018) *'Why the UK Needs a Social Policy on Ageing'*, Journal of Social Policy, Vol 47, No 2, pp 253–273.

WHO. (2012). *Policies and Priority Interventions for Healthy Ageing*. Copenhagen: WHO Regional Office for Europe.

Zaidi, A., & Stanton, D. (2015). *Active Ageing Index: 2014: Analytical Report*. Brussels: UNECE and European Commission Directorate General for Employment, Social Affairs and Inclusion. Retrieved from http://www.southampton.ac.uk/assets/sharepoint/groupsite/Administration/SitePublisher-document-store/Documents/aai_report.pdf

Zaidi, A., & Whitehouse, E. (2009). Should Pension Systems Recognise "Hazardous and Arduous Work"? *OECD Social, Employment and Migration Working Papers, 91*, 1815–199X. https://doi.org/10.1787/1815199X.

Zaidi, A., Gaisor, K., Hofmarcher, M. M., Lelkes, O., Marin, B., & Rodrigues, R. (2013). *Active Ageing Index 2012. Concept, Methodology, and Final Results* (Research Memorandum, Methodology Report). Vienna: European Research Centre.

Zaidi, A., Harper, S., Howse, K., Lamura, G., & Perek-Białas, J. (2018). *Building Evidence for Active Ageing Policies: Active Ageing Index and Its Potential*. Basingstoke: Palgrave Macmillan.

3

Achieving Economic Sustainability in Ageing Societies

Tarmo Johannes Valkonen
and Mikkel Christoffer Barslund

Introduction

Sustainability of public finances has become a part of the everyday political debate. It is most often linked to the fiscal consequences of population ageing but is a broader concept that illustrates whether the current policy leads to continuously increasing public debt in the future. Current policy is a set of rules, which define, for example, the financing of public expenditure by current tax bases and tax rates and the entitlements of the citizens to income transfers and public services. The economic and financial crisis, which started in 2008, revealed large structural budget

T. J. Valkonen (✉)
The Research Institute of the Finnish Economy, Helsinki, Finland
e-mail: tarmo.valkonen@etla.fi; https://www.etla.fi/en/personnel/valkonen-tarmo/

M. C. Barslund
Centre for European Policy Studies, Brussels, Belgium
e-mail: mikkel.barslund@ceps.eu; https://www.ceps.eu/profiles/mikkel-barslund

A. Walker (ed.), *The Future of Ageing in Europe*,
https://doi.org/10.1007/978-981-13-1417-9_3

deficits, increased public debts and lowered the economic growth. As a consequence, the public finance positions of many European Union (EU) countries are markedly weaker just as the challenges of population ageing gather speed. Governments have reacted to the sustainability concerns both by immediate actions, such as tax increases and expenditure cuts, and by structural measures, such as pension reforms (Carone et al. 2016).

It is useful to look at the concept of fiscal sustainability from a broader socio-political perspective. It is just a restriction on policy, not an autonomous societal objective. A fiscal sustainability gap measures the structural change in current fiscal policy necessary to achieve sustainable public finance over the medium to long term. The gap is sometimes used to describe the size of generational conflict. However, there is no reason to remove the part of the gap generated by longer lifetimes of the future generations by increasing taxes or lowering benefits of the current generations (Andersen 2012). But tax rate increases due to other elements of population ageing can slow down economic growth and cause a mismatch between the amounts that the young and future generations pay to the government and how much they receive from it.

Fiscal sustainability is a forward-looking concept. Its numerical estimates are created by combining information on current policies and status of public finances and projections on demographic, economic and health trends. The most relevant economic projections include the rates of productivity growth, real interest and employment. Together with demographic trends they largely determine public spending and the sizes of future tax bases, such as labour and capital income and private consumption. Additional insights can be gained from dynamic general equilibrium models which can account for feedback effects primarily generated by behavioural changes in labour supply and savings of households and investments of firms.

Since the changes in the size of a population and its age structure are generally considered as slow and predictable, population projections have become backbones of the sustainability analysis. Uncertainty related to both the demographic and economic drivers are usually taken into consideration by assuming high and low alternatives to the baseline projections. From the point of view of policy recommendations, this practice has turned out to be problematic, since the errors have been large compared to the range indicated by the alternative scenarios (Keilman et al. 2008).

This chapter discusses uncertainties related to the ongoing trends in demographics and their implications for fiscal sustainability and policies. It also presents new research results on future demographics and simulations on the impacts of policy reforms. Specific attention is given to longer lifetimes.

A Snapshot of Demographic Trends and Economic Effects of Population Ageing

The outcomes of population ageing were discussed already before the Second World War (Dublin 1940). The trends of lower mortality and fertility had already then been visible for some time. Since the provided social security was very limited, the fiscal aspects of ageing were less important. It took almost 50 years before the economic analysis of population ageing was advanced enough and the welfare state entitlements fully developed so that it was possible to analyse comprehensively the changes to be expected in labour markets and public finances.

Recent Trends in Demographics

Fertility rates have been falling almost trend-wise from the beginning of the nineteenth century in Western countries. A notable exception was the baby boom generations born after the Second World War. The long-term trends are not easy to explain and even more difficult to project. As an example of the potential causality dynamics, the Roman Catholic countries had higher than average fertility rates in the 1970s, but now these are among the lowest in Europe. The phenomenon is also not limited to Europe; high-income countries in Asia have some of the lowest fertility rates in the world. The observations from the recent past show some levelling off in the trend decline and even a slight increase in fertility (European Commission 2017). Anyhow, the current number of children per woman of childbearing age is below the replacement fertility rate in all EU countries. A small birth cohort leaves directly almost a century-long imprint on the economy and with the echo effect on the sizes of future birth cohorts, even longer.

Migration flows have recently increased, but the future trends are even more unpredictable than variation in fertility. Efforts to improve living standards and heightened risk of violence are the main drivers of the recent increase in immigration to Europe, but also mobility, in general, has increased due to improved connections, language skills, study opportunities and more liberal immigration policies. The abrupt increase in the number of asylum seekers in Western Europe and the related unrests has changed the political climate so that criteria for asylum have been tightened. But there is still a lot of people living in the nearby areas, who would be eligible if moved.

Life expectancies in Europe have increased from 45–55 years in 1900 to 75–85 years in the 2010s. The speed of increase has varied by country and over time. Official forecasts typically assumed that the decline in mortality would come to a halt. This never happened, and the pessimistic forecasts are partly to blame for the inadequate preparedness of many European societies for the problems of ageing. Even now it is a practice to project that the mortality rate decrease will not continue to be in the future as fast as in the recent past (European Commission 2017).

Declines in mortality first started in the youngest ages with little change in older ages. This led some to conclude that the deaths would concentrate into an ever-narrower range. This has not happened, as illustrated in Fig. 3.1. Declines have gradually started in older ages. It is plausible too that declines will continue to ever higher ages, as resources are devoted to hazards that are characteristic of advanced ages.

The Main Economic Effects of Population Ageing

The economic impacts of permanently low fertility stem from the lower number of children in each household, older age structure and gradual decrease in the population. The lower number of births increases the disposable income per capita available before the offspring leaves home and

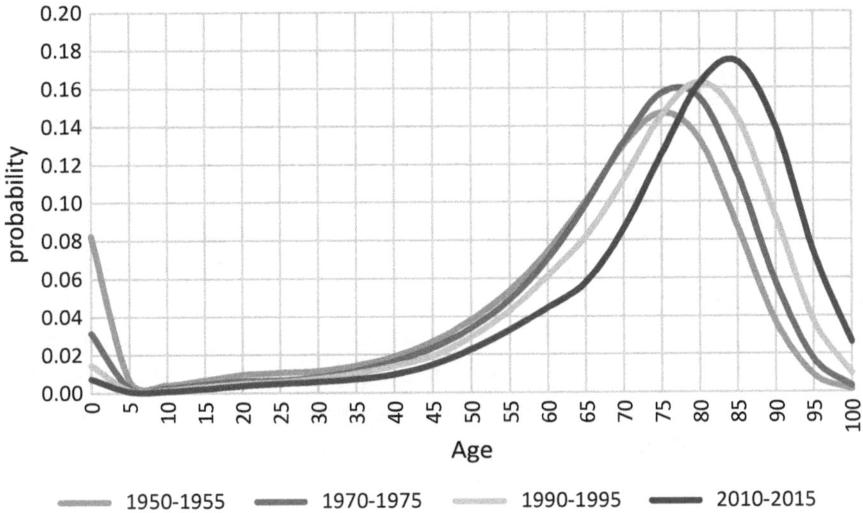

Fig. 3.1 Age distribution of life table deaths during various periods, high-income countries. (Source: UN (2015) and calculations of Nico Keilman)

limits the costs of childhood (or provides a possibility to invest more in health and education of the children). It also reduces the number of future households and magnifies the sizes of received per capita inheritances. Immigrants are typically of young working age and therefore economic effects of immigration are in many respects similar to the ones of a higher fertility if the level of education of migrants is similar to that of native born, except that the investments in care and education during the childhood have taken place in other countries.

The economic consequences of lower mortality rates have changed profoundly in time since the main incidence has moved gradually forward in the life cycle from early childhood to oldest population. The economic results of the lower child mortality were largely similar to the ones due to increase in fertility rates. Higher survival of the young workers had similarities to higher immigration. When the advancements in the survival rates were most visible in the late working years, there was a slight increase in working-age population, but the main consequence was the growth of the retired population. The recent decline in mortality has concentrated in the high-income countries on the late retirement years, where it influences the fiscal and social sustainability of pension schemes and need for health and long-term care services.

From the point of view of individuals, a longer expected life provides many benefits, especially if accompanied by an ability to work longer and engage in active ageing. The economic gains arise mainly from the possibility to earn higher lifetime income, to save more for old age and to consume more. Investments in education and health as young become more profitable. For the society, the economic benefits of longer lives are conditional on the improved working ability, longer working careers and lower age-specific morbidity, which together can largely neutralize the cost pressures. However, if the opportunity is missed and the ratio of the pension years to working years increases continuously, there will be either higher taxes and pension contributions or reduced benefits.

Ageing of the population structure implies that in the goods markets consumption relative to domestic production increases, while both are growing more slowly. In labour markets wages increase as a reaction to the decline in the workforce and higher amount of productive capital per worker used in production. The number of dissaving retirees will increase relative to the number of saving workers, which curtails total saving in the economy. Investment rate falls due to declining number of workers. Consequently, there will be decrease in both the supply and demand for capital, which leaves the direction of the change in the interest rate as an empirical question.

The reactions of the households and firms tend to attenuate the changes in market prices. Higher wages add to investments in human capital, participation in labour markets and working hours, which dampens the effects of fertility decline on the supply of labour. Similarly, any movements in the interest rate will be partly offset by behavioural responses, either from household savings behaviour or from changes in investment.

The economic effects of population ageing interact with public finances. Public pensions, which are typically pay-as-you-go-financed, reduce the need for individuals to prepare for old age by saving and weaken the link between life cycle saving and the interest rate. Moreover, working-age population finances the greater part of many other items of the public expenditure. Therefore, a higher ratio of retirees to workers causes pressures to raise taxes and pension contribution rates, even though the costs of child care and education are smaller. Pensions are taxable, but the corresponding pension contributions are typically deductible in the taxation of wages and profits and their net effect is likely to be negative for the government budget. Higher taxes on labour or consumption weaken the incentives to supply labour and invest in education. Taxing capital income reduces saving for old age and increases thereby interest rates.

International mobility of goods, capital and labour can diminish or enforce the movements in market prices. Higher wages and available jobs tend to attract immigrants. The consequent increase in labour supply restricts the labour market effects of lower fertility. The mobility of capital dampens the influence of the variation in the domestic savings and investment rates on the interest rate. Foreign trade allows access to imported goods and exports of the domestic goods, which limits the variation in prices. However, if the ageing process is simultaneous in other countries, the mobility of goods and factors of production does not have the same cushioning influence on the market prices and may even magnify their variation. Although ageing will affect most middle- and high-income countries in the near future, the process will be faster in Europe relative to the rest of world and, hence, economic factor mobility is likely to counter the impact of ageing in Europe.

Population Ageing, Economic Growth and Secular Stagnation

The ongoing period of low growth and close to zero interest rates on government bonds have provoked interest in the economic effects of population ageing. The state of the economy, called *secular stagnation*, has been explained partly by the global excess supply of savings and diminishing labour force (Eggertsson et al. 2016). The analysis above shows that demographics may indeed contribute to this at least in the phase when the large baby boom generations are still in working age and saving for a longer than expected retirement period. Firms are at the same time reluctant to invest due to expected lack of labour and shrinking growth of consumption, even when the required rate of return on investments is lower. A third potential contribution comes from older age structure of the labour force, which might harm productivity growth (Feyrer 2007; Börsch-Supan and Weiss 2016).

There are still major weaknesses in this part of the secular stagnation story because in such early ageing countries as Japan and Germany, a large share of the baby boomers has already retired. Still, investment rates are not falling and current accounts are in surplus, showing that saving continues to be larger than investments. Acemoglu and Restrepo (2017) show that countries experiencing more rapid ageing have grown more in recent decades. They suggest that these countries adopt more rapidly automation technologies.

Another major counteracting force for the gloomiest predictions of the secular stagnation studies is the strong improvement in the employment rates of 55–64-year-old workers. There are many possible explanations for this, such as a shift in skill composition of the labour force, more broad-based participation of women in labour markets, social security reforms (Maestas and Zissimopoulos 2010) and improved health. The two first mentioned are likely to lose ground in the coming years, but many of the retirement reforms will only show their full effect in the future due to the gradual implementation and due to the links introduced between pension age and life expectancy.

Also, most of the multi-country simulation models indicate that the decline in interest rates or productivity growth due to population ageing is likely to be rather limited and realize slowly (Börsch-Supan et al. 2006). Some simulations even suggest that world interest rates may increase (Fehr et al. 2013).

New Results on Demographics and Economics of Population Ageing

Traditional Measures of Population Ageing Are Problematic

Population ageing has traditionally been assessed by looking at the old-age dependency ratio (OADR), which measures the ratio of pensioners per working-age adult. This measure leaves out the large group of young people aged up to 20 years old. The main reason for this distinction is that pensioners affect public finances substantially more directly than young people due to pension benefits and much larger health care costs. However, falling fertility rates has been the other major driver of population ageing in most European countries. This implies that focusing only on the OADR may underestimate the public finance challenges facing many countries because in the previous two decades a falling number of young people has aided the public finance adjustment process.

On the other hand, part of the ageing challenge is directly due to increased longevity, and therefore at least seems desirable from a human and societal point of view. Several authors have proposed the so-called prospective measures of ageing which in various ways account for increasing longevity (e.g. Sanderson and Scherbov 2013; Spijker and MacInnes 2013).

The study of Barslund and Werder (2016b) provides a comprehensive overview of the different measures, in particular, prospective measures of ageing, available to quantify the challenges stemming from population ageing, and which are designed to take increases in longevity into account. The study is also concerned with economic dependency ratios, that is, number of inactive to labour market active persons. This is arguably a more relevant measure for summarizing the economic challenges related to ageing.

They find that prospective indicators of ageing reveal the challenge of population ageing to be less immense than traditional chronological measures suggest. However, the study emphasizes that the relevance of prospective indicators rests on the premise of policy changes emerging according to the changing age structure of the population. Second, further substantial increases in the length of working lives are necessary to maintain current economic dependency ratios. Progress, although uneven across countries, in increasing working lives has already been made since 2000. Due to higher labour force participation rates among especially the 50–74-year-olds, the economic dependency ratio declined in the period 2000–2013. Part of this improvement stems from changes in the educational composition of the population with fewer low-skilled workers who tend to retire earlier. While better education of older workers will continue to improve prospectively, the direct impact on the scale of the ageing challenge will be more limited.

Number of Older People Living Alone Grows Strongly

A number of other trends interact with the changes in the population structure. One of them is the changing demographic structure of households. The number of people living in various types of households has implications for both policy and living standards. For example, the increasing number of older people living alone adds to the need for housing, private and social insurance and health and long-term care services.

Policy planning is complicated by the marked uncertainty around future household structures. Keilman (2015) assesses this uncertainty by creating probabilistic household projections for the years 2011–2041. Using population registers of Denmark, Finland and the Netherlands, Keilman (2015) estimates the uncertainty parameters from time series data. For the two other studied countries, Germany and Norway, the uncertainty parameters were borrowed from the other three countries.

His findings show that on current trends, we will see more and smaller households, often driven by increasing numbers of persons who live alone. The number of 80+ people in one-person households will be 2–4 times higher during next 30 years in the studied countries. The uncertainty related to this outcome is not very large. The percentage change is bigger for men, but absolute change is far larger for women. Figure 3.2 illustrates the projections for men and women aged 80+ living alone in Germany. For instance, there is an 80 per cent chance that in 2041 there will be between 0.9 and 1.8 million men who live alone, with the best guess of 1.3 million. The numbers will jump already during the current decade and again in the 2030s when the baby boomers reach the age limit of 80. This age group is of considerable interest for policymakers. Although the health condition of the oldest old may improve in the years to come, many of these will be in need of formal and informal care.

The count of older people living alone is related both to voluntary decisions, such as divorces and remarriages, and to the unfortunate

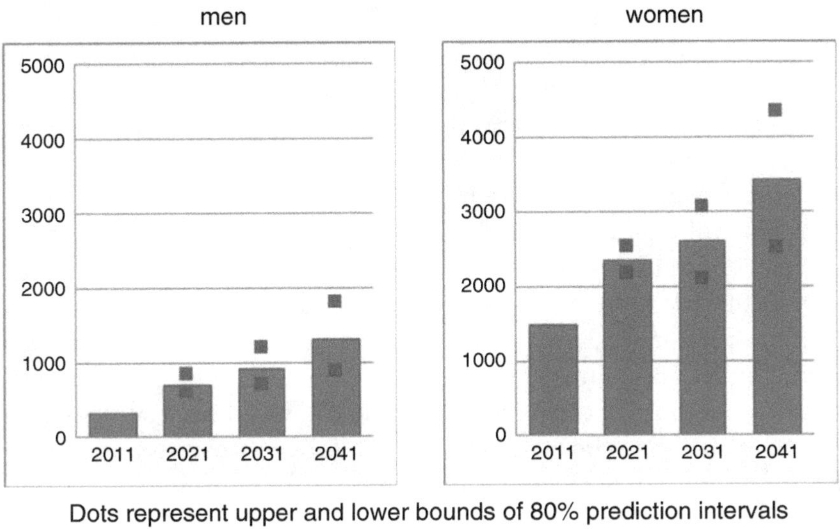

Fig. 3.2 Germany: numbers of men and women aged 80+ who live alone (in 1000s). Census (2011) and predicted. (Source: Keilman 2015)

cases where the spouse is institutionalized or dies. Male mortality has almost universally been higher than female mortality, which is also the main factor behind the gender difference in the numbers in Fig. 3.2. As the economic and social roles of the two genders have become more similar in Europe, with respect to work, child rearing, leisure activities and other aspects of life style, many expect that male mortality will catch up with female mortality. This convergence has been assumed, for example, in the population projections of Eurostat (see European Commission 2017).

Alho (2016) analysed the historical mortality convergence in 23 European countries using data from Human Mortality Database. It turned out that the countries can be classified into three broad groups: (1) former socialist countries, in which male mortality is clearly higher than female mortality; (2) countries around the North Sea and the Atlantic that might be characterized as egalitarian from the perspective of gender equality with a sharp increase in the gap between male and female life expectancy after World War II, and with a subsequent rapid, but partial recovery; and (3) the remaining countries from Finland to Central

Europe and the Mediterranean, in which divergence in expected years started to increase a century ago, with an acceleration after World War II, and a more modest recovery. In contrast to the egalitarian group, these countries might be characterized as traditional from the perspective of gender equality.

Figure 3.3 shows that there is a high correlation between the relative probability of males and females surviving from birth to age 65 and the ratio of expected years of males relative to females, at 65. The data is from years 2009–2014. In the former socialist countries, the male disadvantages are the highest. Alho (2016) also reveals a considerable diversity from country to country, even when the broad groupings are applied. Therefore, even if the convergence of male and female mortalities is a real possibility, it cannot be taken for granted, for every country, in forecasting, for example. Considerable uncertainty will remain.

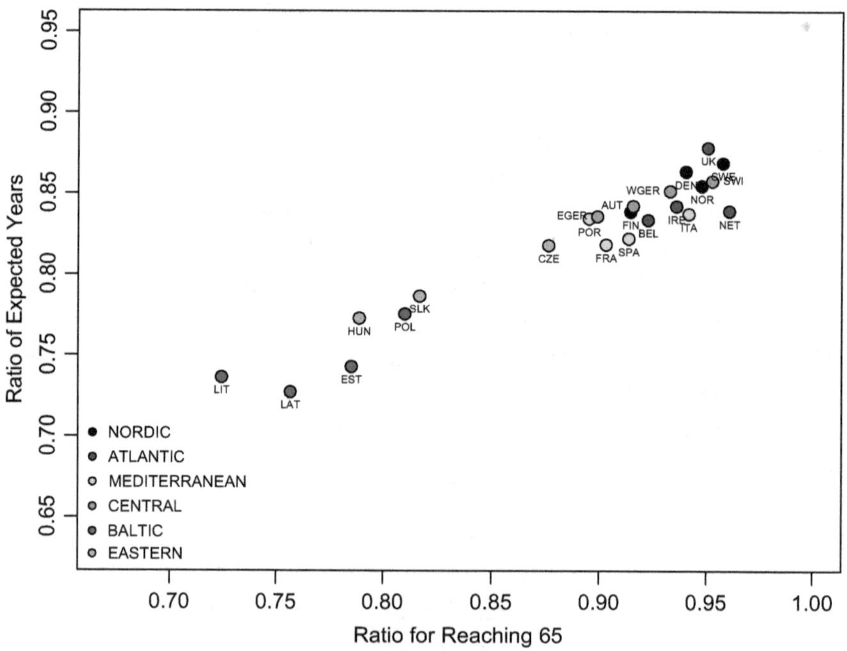

Fig. 3.3 Ratio of expected years of life in ages 65–99, as a function of the ratio of probabilities of surviving to 65. (Source: Alho 2016)

Reducing Fiscal Costs of Population Ageing

Empirical studies on the economic consequences of population ageing can roughly be divided into three categories depending on their methodology. Macroeconometric models and numerical overlapping generations (OLG) models (Fehr et al. 2013) imitate the behavioural reactions of the economy to the change in demographic structure. Numerical accounting models, such as National Transfer Accounts, are based on demographic projections and extensions of the national accounts with age-specific disaggregation of the public and private transfers (Lee 2016). Third, there are statistical estimations of relationships of interest, such as the historical influence of the share of the elderly population on economic growth (Maestas et al. 2016).

Two recent studies have used the links between longer lifetimes, longer working lives and better health as inputs in numerical models to provide a profound analysis on fiscal sustainability.

Barslund and Werder (2016a) use National Transfer Accounts of seven countries to project the quantitative impact of ageing on public finances until 2040 and compare it to projected OADRs. They simulate the public finance impact of changes in three key indicators related to the policy responses to population ageing: net immigration, healthy ageing and longer working lives. Healthy ageing is approximated by linking age-specific public health transfers to changes in mortality. Correspondingly, longer working lives are implemented in the simulations by linking labour market participation rates to changes in mortality.

It turns out that the simple OADR overestimates the future public finance challenges faced by the countries studied. In some case, as for Austria, Finland and Hungary, this is significant. Second, healthy ageing—keeping health transfers constant for a given mortality rate—has a modest effect except in the case of Sweden where it is important. Third, the long-run effect of immigration is well captured by the OADR measure if the common assumption of immigrants being similar to natives is maintained, while the immediate to short-term impact of immigration tends to be overstated. Finally, increasing the length of working lives is central in addressing the public finance challenge of ageing. If government transfers are linked to mortality, extending the average length of working lives with around 3–4 years over the next 25 years—roughly equivalent to the gain in life expectancy—severely limits the impact of ageing on public transfers.

The study of Lassila and Valkonen (2017) investigates whether longer work careers can bring sufficient increases to tax revenues to offset the negative effects of growing pension, health and long-term care expenditure that longer lifetimes cause. Uncertainty in future mortalities is considered using 500 realizations of stochastic mortality simulations as inputs in an open economy version of the Auerbach-Kotlikoff numerical overlapping generations model. The model is well suited for the analysis of the economic effects of population ageing, since the population structure of each period can be described with a lot of precision. The model dynamics stem from the reactions of the households and firms to changes in demographic variables, market prices and economic policy (see Lassila 2015).

Health and long-term care expenditures are analysed using two options. In the naive specification, the per capita expenditure is purely age-related. In the more developed version, the need for services is linked to both age and proximity of death. These links are specified using individual register data from Finland.

Policy analysis concerns the rules of the pension system. The first work career option is a simple reference outcome where the average effective retirement age follows a fixed path. According to the estimates of Määttänen (2014), adding an additional three years to the life expectancy of a 30-year old would extend working lives by six months, assuming that any health problems are likewise postponed by three years, even if pension rules were left unchanged. The second specification describes the Finnish earnings-related pension system applied in 2005–2016, where there was an adjustment of monthly benefits that depends on cohort-wise life expectancy. It was hoped that this adjustment would lead to postponed retirement, but outcomes were disappointing.

The third specification in Lassila and Valkonen (2017) refers to a reform where also cohort-wise pension eligibility ages are linked to life expectancy (Andersen et al. 2014 provides design options for this policy). It resembles the reform implemented in Finland 2017. In the reform, the pensionable age adjusts every year to changes in mortality so that it divides the expectancy for the time lived as an adult into working lives and retirement years at the same ratio (roughly 2:1). Based on the results by Määttänen (2014), increasing the eligibility ages of the pensionable age, the unemployment pathway and the part-time pension by two years would extend working lives by seven months. This estimate has been used in the model by Lassila and Valkonen (2017) in such a way that the change in life expectancy automatically affects the length of working lives in accordance with the 2:1 ratio depicted.

For each of the three scenarios, the resulting fiscal sustainability gap (equivalent to the S2 indicator used by the European Commission) can be calculated for different future mortality scenarios (Fig. 3.4). These calculations use the proximity-to-death approach to estimate future health and long-term care expenditure. The top set of dots (blue) show

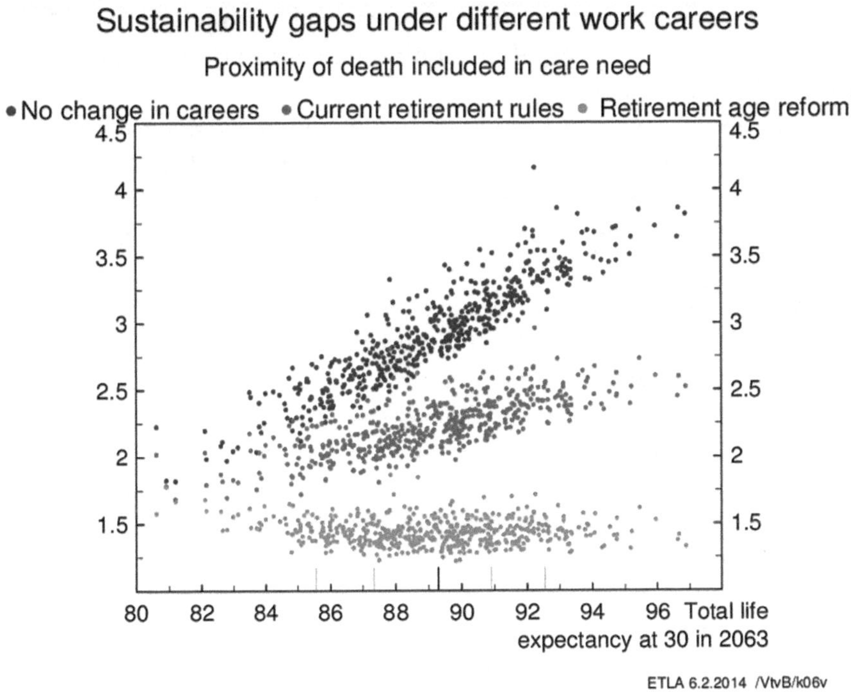

Fig. 3.4 Total life expectancy at 30 in 2063 and sustainability gaps under different working lives. (Source: Lassila and Valkonen 2017)

outcomes of a typical sustainability calculation, where the labour supply and saving reactions of the individuals to longer lifetimes are missing. If these reactions are considered, the sustainability gaps are smaller, as shown by the middle dots (red). But it requires a retirement age reform as specified above to insulate the fiscal sustainability from mortality uncertainty, as illustrated by the bottom set of dots (green).

The results show that it is quite possible that longer working lives bring sufficient increases to tax revenues to offset the effects of growing expenditure that longer lifetimes cause. The outcome requires policies that link eligibility ages for pensions to life expectancy and fully utilize the improvement in health.

Lassila and Valkonen (2014) show that with a yearly net immigration of 17,000 persons (0.3 per cent of the Finnish population), fiscal sustainability gap could be reduced by one percentage point if the employment rates of the immigrants were equal to the ones of the original population. But adjusting yearly immigration is impractical and a politically unfeasible way of neutralizing the effects of mortality uncertainty on the sustainability of public finances.

There are also other possibilities to increase the resilience of public finances to longer lifetimes. In pay-as-you-go financed defined benefit pension schemes, where pension contributions are tax deductible and pensions taxable (exempt-exempt-taxed (EET) regime), governments bear part of the pension expenditure risk due to uncertain longevity. Higher life expectancy augments pension expenditure and increases the contribution rate. At the same time, both the tax deductions of the contributions and the tax revenues generated by taxing pensions increase. The net effect for the government budget is negative since the average tax rate of pensions is typically lower than the income tax rates used when deductions are made. Another reason for the loss of government is that in pay-as-you-go schemes the earned pension is typically low compared to the paid lifetime contributions (after the scheme has passed the implementation phase).

In taxed-exempt-exempt (TEE) tax regime, where contributions are not deductible, but pensions are tax-free, higher pension expenditure influences neither tax deductions nor tax revenues collected from pensions. There is no direct link between the income tax revenues and pension expenditure.

Lassila and Valkonen (2018) simulate the effects of the tax regime shift from EET to TEE in Finland's earnings-related, first-pillar pension scheme. Uncertainty in longevity is considered using 500 sample paths from a stochastic mortality projection in the used numerical OLG model. Abolishing deductibility of pension contributions has strong indirect effects on tax bases. Since the Finnish earnings-related pension scheme is mainly financed by employers' contributions, firms' reaction to the increase in labour costs dominates the outcomes. Wage rates and demand for labour fall, which decreases income and consumption. The total tax rate increases due to tightened taxation of profits and labour income. Moreover, lower wages decrease public sector wages and income transfers indexed to wages. Therefore, the tax regime shift improves public finances the more the higher is longevity, but at the expense of less efficient tax structure and lower consumption. These results were compared to a case where the pension scheme adjusts to higher longevity by linking the pension benefits and pension age to life expectancy. In this automatically adjusting pension scheme, the tax regime shift is not so efficient, but more importantly, not so necessary either.

A fourth policy simulation study by Fehr et al. (2015) utilizes the multi-region OLG model constructed in Fehr et al. (2014). The model builds on the overlapping generation structure from Auerbach and Kotlikoff (1987) but offers two major extensions. First, it provides a multi-country set-up to quantify the international spill-over effects of policies and, second, it features a disaggregated production and consumption structure to analyse the sectoral growth and trade effects. In the baseline path, US gross domestic product (GDP) expands by a factor of 2.45; GDP only roughly doubles in both Western Europe and North East Asia through 2060. These differences mainly reflect demographic differences, particularly the absolute population decline in Europe and North East Asia. However, due to the major productivity increases assumed to arise in China and India, they become the largest economies in the next decades.

The simulated policy reforms aimed to dampen the future economic effects of population ageing are a move towards consumption taxation, an elimination of capital income taxation, a significant debt reduction or a pension funding reform in selected Western economies. The four policies induce a very similar macroeconomic adjustment process during the transition to the new steady state equilibrium. In the first phase after the reform, there is either an increase in capital supply (due to higher savings) or a reduction in capital demand (due to reduced debt) in the reforming region. The increased availability of capital induces a shift in activities from all other sectors towards the investment good sector.

In the medium run, households in the reforming regions experience positive income effects which will reduce their labour supply. The reduced labour supply leads to a capital outflow towards non-reforming regions and a shift of production and resources away from investment goods towards the housing sector in the reforming region. Consequently, non-reforming countries first benefit from capital inflows and potentially an additional positive terms of trade effects which ultimately translates into significant welfare gains for their future cohorts. It turns out that mainly the North East Asia region benefits from reform policies in Europe and the United States.

The most preferred policy turns out to be a debt reduction financed by an increase in wage taxes and consumption taxes since it has only modest intersectoral effects and international spill-overs while at the same time the intragenerational redistribution favours low-skilled households.

Fiscal Sustainability and Redistribution

People with low education have on average lower income, higher unemployment rate and shorter life expectancy. Improving fiscal sustainability by longer working lives is therefore often seen as problematic for equity reasons. Auerbach et al. (2017) show, however, that it is the increasing life expectancy gap that generates regressivity in the pension systems, not the policy of increasing retirement ages.

Määttänen (2015) analyses the distributional effects of pension reforms that aim at extending working lives. The study uses a numerical life cycle model of labour supply decisions that incorporates a detailed description of several elements of the Finnish social security system. The model accounts for systematic and substantial differences in average age-wage profiles, disability risk and mortality across groups that differ by gender and education as well as within group heterogeneity resulting from idiosyncratic uncertainty.

The studied reform restricts early retirement and increases the eligibility age for old-age pensions. It turns out that such a reform is likely to decrease the lifetime benefits of the highly educated individuals the most and to reduce lifetime welfare inequalities. The main reason for this relates to the progressivity of the overall social security system. Those with the lowest lifetime income are usually long-term disability pension retirees. Raising the earliest eligibility age for old-age pension does not decrease their lifetime benefits. Low-wage workers are also relatively well protected against layoff risk at old age by progressive unemployment insurance.

On the other hand, the highly educated individuals, who tend to retire much later than others, benefit from higher accrual rates starting at the eligibility age for the old-age pension. By increasing the eligibility age for old-age pensions, those high accrual rates can also be postponed. These results apply to the Finnish social security system only, but it seems likely that progressive unemployment insurance and disability pensions can be used to protect many relatively poor individuals against the possible adverse effects of pension reforms in other countries as well.

There is a related discussion that prefers linking retirement age to healthy life expectancy (HLE) instead of life expectancy (AGE Platform 2016). One justification is obvious: the prerequisite for longer working lives is better health near the retirement age. This is an interesting suggestion that requires more research. One of the problems in this proposal is an imprecise measurement of health. For example, the Healthy Life Years (HLY) measure seems to vary very much between countries and in time. Another problem is that there is little correlation between these HLY and employment rates of the elderly workers, which suggests that HLY is not the factor that restricts higher employment rates in many countries.

Figure 3.5 shows the low correlation. With the same value of HLE at age 50, the employment rate of 55–64-year-old males varies 20–30 percentage points. Only in some Nordic countries, such as Iceland, Norway and Sweden, and in Switzerland, high values of both indicators are achieved at the same time. There are also indications that while the years

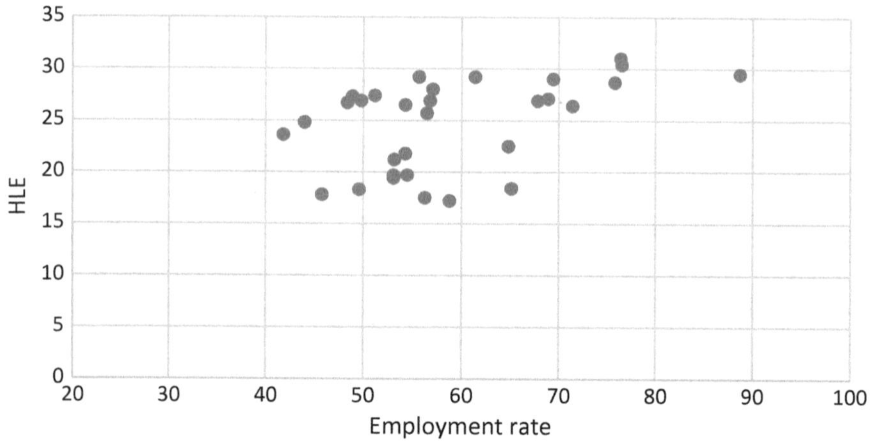

Fig. 3.5 Employment rates, age 55–64 and HLE at age 50 (self-perceived health) in 31 European countries in 2014, males. (Source: Eurostat)

lived with chronic conditions are increasing with higher longevity, the incidence of severe disabilities during the working age has reduced.

Another reason for linking the retirement age with health is the perception that health outcomes at older ages are diverging among socioeconomic groups. However, as with HLY, more research is warranted to examine the trend in health inequality among socioeconomic groups. Two recent studies suggest that trends in health inequality may not be a big concern (Mackenbach et al. 2016; Barslund and Ludolph Forthcoming).

Policy Conclusions

Common recommendations aiming to restore fiscal sustainability in the ageing EU economies must be formulated at a general level for several reasons. Populations in various countries are at different stages of ageing and also the profundity of the future age structure changes varies between the countries. Moreover, the public finance consequences of a given age structure change vary, depending on the details of the tax and social security systems. Many of the already decided reforms are still in their implementation phase.

A second general observation is that one-off measures that are aimed to abolish a given sustainability gap are bound to be insufficient, or overblown since the future is uncertain. An example is a one-time increase in the statutory retirement age. It can later be repeated, if needed, or cancelled, if considered excessive, but with a non-negligible amount of re-negotiations and implementation time, which may postpone the adjustment for a very long time. In the worst-case scenario, the necessary fiscal sustainability adjustments are postponed due to political and administrative considerations so long that the needed policy changes are abrupt, creating problems for intra- and intergenerational redistribution. Therefore, more focus should be put on automatic rules that increase the resilience of public finances. Of course, the social sustainability of the rules must be checked beforehand.

The new research results discussed in this chapter give scientific reasons for some more detailed policy conclusions, too. The burgeoning numbers of older people living alone require a broad set of policies. Individuals should be informed about the increased economic risks of living alone and for longer than expected. Provisions for old age require both private saving and use of insurance products. Need for smaller apartments, as well as the demand for community housing increase. People living in one-person households also gain from social insurance more and use public health and long-term care services more frequently.

Measurement of population ageing needs more attention. The average age of retirement has increased already several decades and more is to be expected with longer lifetimes, retirement reforms and better education. The cut-off age of 65 years will be less and less relevant to describe participation in labour markets. This should be reflected in calculations of demographic and economic dependency ratios and employment rates. Alternative ageing measures indicate that the challenge of population ageing is less immense than traditional chronological measures would suggest, but also that substantial increases in the length of working lives are necessary to maintain current economic dependency ratios and to address the public finance challenges of ageing.

Later retirement is an obvious response to longer lifetimes. The current practice of fixing the statutory eligibility ages is a leftover from times when life expectancies were low and last wages determined replacement rates. Flexible old-age retirement windows, which move to keep the ratio of working and retirement years unchanged when longevity increases, are needed. Flexibility helps individuals to adjust their retirement to weakened working capacity, discontinuous working lives, employment prospects and preferences for leisure. Tying the age limits of the retirement window to life expectancy helps to keep the pension schemes fiscally resilient.

To preserve social sustainability, the earnings-related pension schemes must be complemented with basic pensions, disability pensions and unemployment insurance. A well-functioning disability pension scheme is likely to be a more precise instrument for concerns of individually varying health and working capacity near the end of the working life than a link between the overall retirement age and the less accurately measured average healthy life years.

References

Acemoglu, D., & Restrepo, P. (2017). Secular Stagnation? The Effect of Aging on Economic Growth in the Age of Automation. *American Economic Review, 107*(5), 174–179.

Age Platform Europe. (2016, December). *Building an Age-Friendly Europe with the European Pillar of Social Rights*. AGE Platform Europe Policy Statement, Brussels.

Alho, J. M. (2016). *Descriptive Findings on the Convergence of Female and Male Mortality in Europe* (Etla Working Papers, 40).

Andersen, T. M. (2012). *Fiscal Sustainability and Fiscal Policy Targets*. (Economics Working Papers 2012–15, Department of Economics and Business Economics, Aarhus University).

Andersen, T. M., Määttänen, N., & Valkonen, T. (2014). Pension Reforms: Longevity and Retirement. In T. Valkonen & V. Vihriälä (Eds.), *The Nordic Model – Challenged but Capable of Reform*. Nordic Council of Ministers. Copenhagen.

Auerbach, A. J., & Kotlikoff, L. J. (1987). *Dynamic Fiscal Policy*. Cambridge: Cambridge University Press.

Auerbach, A. J., Charles, K. K., Coile, C. C., et al. (2017). How the Growing Gap in Life Expectancy May Affect Retirement Benefits and Reforms. *The Geneva Papers on Risk and Insurance – Issues and Practice, 42*(3), 475–499.

Barslund, M., & Ludolph, L. (Forthcoming). Trends in Health Inequality Among Older People in European Countries.
Barslund, M., & von Werder, M. (2016a). Measuring Dependency Ratios Using National Transfer Accounts. *Vienna Yearbook of Population Research, 14*, 155–185.
Barslund, M., & von Werder, M. (2016b, February). *Measuring Ageing and the Need for Longer Working Lives in the EU* (CEPS Working Document, 417).
Börsch-Supan, A., & Weiss, M. (2016). Productivity and Age: Evidence from Work Teams at the Assembly Line. *The Journal of Economics of Ageing, 7*, 30–42.
Börsch-Supan, A. H., Weiss, M., Ludwig, A., & Winter, J. K. (2006). Ageing, Pension Reform and Capital Flows: A Multi-Country Simulation Model. *Economia, 73*(292), 625–658.
Carone, G. P., Ecklefeldt, L., Giamboni, V. L., Laine, S., & Pamies, S. (2016). *Pension Reforms in the EU since the Early 2000s: Achievements and Challenges Ahead*. Brussels: European Commission, Discussion Papers 42.
Dublin, L. J. (1940). Problems Arising from an Ageing Population. *Nature*, 146–566.
Eggertsson, G., Mehrotra, N., & Summers, L. (2016). Secular Stagnation in the Open Economy. *American Economics Review, Papers and Proceedings, 106*(5), 503–507.
European Commission. (2017). *The 2018 Ageing Report: Underlying Assumptions and Projection Methodologies*. Institutional paper 065, Brussels.
Fehr, H., Jokisch, S., Kallweit, M., Kindermann, F., & Kotilikoff, L. J. (2013). Generational Policy and Aging in Closed and Open Dynamic General Equilibrium Models. In P. B. Dixon & D. W. Jogenson (Eds.), *Handbook of Computable General Equilibrium Modeling* (pp. 1719–1800). North Holland: Elsevier.
Fehr, H., Jokisch, S., Kindermann, F., & Kotlikoff, L. J. (2014). *Description of Extended Simulation Model*. Mimeo, Project MOPACT. Retrieved from http://mopact.group.shef.ac.uk/wp-content/uploads/2013/10/D2.1-Economic-consequences-extended-simulation-model2.pdf
Fehr, H., Jokisch, S., & Kotlikoff, L. J. (2015). *Report on the Impact of Demographic Change on Production* Structures *and Trade*. Mimeo, Project MOPACT. Retrieved from http://mopact.group.shef.ac.uk/wp-content/uploads/2013/10/D2.2-Economic-consequences-Impact-on-production-trade.pdf
Feyrer, J. (2007). Demographics and Productivity. *The Review of Economics and Statistics, 89*(1), 100–109.
Keilman, N. (2015). *Probabilistic Household Forecasts for Five Countries in Europe*. Mimeo, Project MOPACT. Retrieved from http://mopact.group.

shef.ac.uk/wp-content/uploads/2013/10/D2.5-Probabilistic-household-forecasts-for-five-countries-in-Europe.pdf

Keilman, N., Cruijsen, H. & Alho, J. M. (2008). Changing View of Future Demographic Trends. In Alho, J. M., Hougaard Jensen, S. E. & Lassila, J. (Eds.), *Uncertain Demographics and Fiscal Sustainability*. Cambridge: Cambridge University Press.

Lassila, J. (2015). *Quantifying the Influence of Demographic Transition on the Public Finances in Finland*. Mimeo, Project MOPACT. Retrieved from http://mopact.group.shef.ac.uk/wp-content/uploads/2013/10/Quantifying-the-influence-of-demographic-transition-on-public-finances-in-Finland-1.pdf

Lassila, J., & Valkonen, T. (2014). Public Finances, Markets and the Health and Long-Term Care Services. In T. Valkonen & V. Vihriälä (Eds.), *The Nordic Model – Challenged but Capable of Reform*. Nordic Council of Ministers. Copenhagen.

Lassila, J., & Valkonen, T. (2017). Longevity, Working Lives and Public Finances. *Contemporary Economic Policy, 36*, 467–482. https://doi.org/10.1111/coep.12262.

Lassila, J., & Valkonen, T. (2018). Longevity Risk and Taxation of Public Pensions: Finland. In R. Holzmann & J. Piggott (Eds.), *The Taxation of Pensions*. MIT Press.

Lee, R. (2016). Macroeconomics, Ageing, and Growth. In J. Piggott & A. Woodland (Eds.), *Handbook of the Economics of Population Ageing* (Vol. 1, pp. 59–118). North Holland: Elsevier.

Määttänen, N. (2014). Evaluation of Alternative Pension Policy Reforms Based on a Stochastic Life Cycle Model. In J. Lassila, N. Määttänen, & T. Valkonen (Eds.), *Linking Retirement Age to Life Expectancy – What Happens to Working Lives and Income Distribution?* Helsinki: Finnish Centre for Pensions.

Määttänen, N. (2015). *Report on the Impact of Measures to Promote Labour Supply on the Welfare of Older People*. Mimeo, Project MoPAct. Retrieved from http://mopact.group.shef.ac.uk/wp-content/uploads/2013/10/Report-on-the-impact-of-measures-to-promote-labour-supply-on-the-welfare-of-older-people.pdf

Mackenbach, J. P., Kulhánová, I., Artnik, B., Bopp, M., Borrell, C., Clemens, T., & Martikainen, P. (2016). Changes in Mortality Inequalities Over Two Decades: Register Based Study of European Countries. *British Medical Journal, 353*, 1732.

Maestas, N., & Zissimopoulos, J. (2010). How Longer Work Lives Ease the Crunch of Population Ageing. *Journal of Economic Perspectives, 24*(1), 139–160.

Maestas, N., Mullen, K. J., & Powell, D. (2016). *The Effect of Population Ageing on Economic Growth, the Labour Force and Productivity* (NBER Working Papers). National Bureau of Economic Research, Inc, 22452.

Sanderson, W. C., & Scherbov, S. (2013). The Characteristics Approach to the Measurement of Population Ageing. *Population and Development Review, 39*, 673–685. Retrieved from. https://doi.org/10.1111/j.1728-4457.2013.00633.x.

Spijker, J., & MacInnes, J. (2013). Population Ageing: The Timebomb That Isn't? *British Medical Journal, 347*, f6598. https://doi.org/10.1136/bmj.f6598.

UN. (2015). *World Population Prospects: The 2015 Revision. United Nations*, Population Division. Retrieved from https://esa.un.org/unpd/wpp/

4

Effective and Sustainable Private Pensions

Theo Nijman and Arthur van Soest

Introduction

In many countries around the globe, income provision during retirement is more and more dependent on funded defined contribution (DC) pensions. In the European Union (EU), the European Insurance and Occupational Pensions Authority (EIOPA) advocates the so-called Pan European Personal Pensions (PEPPs) which are to provide a standardized DC pension product through a pan European regime. One of the policy aims of the EU here is to stimulate a better balance between unfunded (pay-as-you go [PAYG]) and funded pensions. In other countries (e.g. the United States, the United Kingdom, and the Netherlands), the traditional defined benefit (DB) pension systems are under pressure and are more and more replaced by DC pensions.

T. Nijman (✉) • A. van Soest
Tilburg University/Netspar, Tilburg, The Netherlands
e-mail: nyman@uvt.nl; A.H.O.vanSoest@uvt.nl; https://www.tilburguniversity.edu/webwijs/show/nyman.htm; https://www.tilburguniversity.edu/webwijs/show/a.h.o.vansoest-1.htm

A. Walker (ed.), *The Future of Ageing in Europe*,
https://doi.org/10.1007/978-981-13-1417-9_4

In DC pensions, capital is invested during the working years in the so-called accumulation phase. In the decumulation phase, that is, during retirement, the capital is gradually withdrawn. DC pensions therefore require choices how much to contribute to the pension system, how much to drawdown, how to invest the capital, and which risks to insure in the pension system. Longevity risk is one of the most important risk factors for pension income. If longevity risk is insured, for example, using annuity contracts, a lifelong income for the participant is obtained. One of the main drawbacks of full insurance with annuities is that there will be no bequest. Likewise, whether or not to buy insurance to have income for a partner if an individual passes away is an important insurance decision.

This chapter summarizes the work done in the Mobilising the Potential of Active Ageing in Europe (MOPACT) project on adequate choices for contributions rates, drawdown rates, investment, and insurance in DC pensions. Such choices can be purely made by individuals or fully by trustees of a pension plan or by the regulator. In reality, however, most choices are neither made purely by the individual nor fully mandated. Often the regulator restricts the choice set, subsequently trustees of a pension fund restrict the choice set further and provide default solutions for individuals who do not actively choose, and finally, individuals have some remaining freedom of choice, because they have the best information on their own preferences and characteristics. Because adequate decision-making is key in the provision of adequate DC pensions, the MOPACT project also spent considerable effort to better understand how people make pension-related decisions. The recent literature has convincingly demonstrated that individuals often do not make the decisions that are in their own best interest, particularly when it comes to complicated decisions with long-term consequences and substantial uncertainty. This raises the question how optimal decision-making can be stimulated. Recent insights from psychology and behavioral economics emphasize the role of individuals' financial and pension literacy and capabilities of pension planning, but also pension awareness and involvement in pension decisions. Pension communication and information provision as well as the framing of the decision process play an important role. In particular, tools from choice architecture (defaults, message framing, timing, etc.) can be used to 'nudge' people toward making adequate decisions.

The Key Challenges

Funded DC pensions currently provide two key challenges to regulators, supervisors, and trustees of pension providers, as well as to individuals:

1. How to optimally contribute to, drawdown from and invest DC pension capital, and how to insure the biometric risks (risk of getting old, risk of losing the partner), given assumptions on individual preferences and characteristics?
2. How to structure the decision processes (including legislation and supervision) accounting for the heterogeneity in preferences, circumstances, involvement and financial literacy, and the preference for freedom of choice?

The first key challenge becomes more and more urgent as DC pensions mature. While in the past DC pensions were important primarily in the accumulation phase it has now become a substantial part of retirement income for many. As we will discuss further in the next section, it is in particular the decumulation phase that is under researched. Note also that the PEPPs fully neglect the decumulation phase. More attention here is the more pressing because of the low interest rate environment. Pension products can easily be constructed and understood if they provide income guarantees, and such guarantees are highly preferred by individuals if one does not refer to cost of the guarantees. Low interest rates imply though that guarantees are very costly and come with riskless investing during the retirement period (of often more than 20 years) which is unattractive unless one is very risk-averse. This has probably been one of the reasons why mandatory guaranteed annuities have been abolished in the United Kingdom. In other countries (Denmark, the Netherlands) where lifelong income is still one of the cornerstones of the pension system, new products have emerged that provide limited income guarantees.

The pension reforms that have been or will be introduced in many countries typically give more responsibility for an adequate standard of living after retirement to individuals and their households. Whereas the traditional occupational DB pension at a mandatory retirement age automatically more or less guaranteed a certain pension income, the shift to DC pensions with more flexibility in retirement age, investment strategy, and decumulation pattern, comes with choice responsibility for individuals and private households. In Italy and Germany, for example, the reformed public pension system is much less generous than the old one, and workers themselves are responsible for supplementing their retirement savings with (often tax-favored) voluntarily savings. More choice also becomes more common at later lifecycle stages. In the Netherlands, flexible and gradual retirement opportunities have become more common, with actuarially neutral trade-offs between earlier retirement or a higher pension income. After retirement, occupational DB pensions nowadays often allow for a non-constant annuity (such as a high-low construction) and there are plans to make it possible to take up part of the mandatory occupational pension savings as a lump sum instead of an annuity. For DC pensions, the option to invest in risky assets implying a variable annuity has been introduced. All these changes imply that much more than in the past, important pension-related choices have become the responsibility of the individuals themselves rather than their pension fund, their employer, or the government.

On the one hand, increasing the choice opportunities for individuals is a positive development, since it allows them to make choices that fit optimally with their own personal situation, their preferences, and their expectations and ambitions for the future. On the other hand, more choice is not necessarily better if individuals do not make optimal decisions. There is abundant evidence that, particularly when it comes to complex financial problems such as those related to pensions, many individuals do not make the decisions that are in their own best interest. In other words, the *homo economicus* hardly exists, and instead of maximizing expected utility under rational expectations, many individuals decide on the basis of rules of thumb and defaults. Their decisions are strongly influenced by framing and contextual cues that do not affect the actual consequences of the choices or the optimal choice.

This naturally leads to the question how pension funds, the government, employers, and perhaps other parties involved can help individuals to make decisions that are in their own best interest (in the sense of maximizing their expected utility over the lifecycle). The first important aspect of this is financial literacy. An active literature has shown that many people have very low financial literacy. Moreover, lack of financial literacy may lead to less retirement planning. Stimulating financial literacy may therefore improve retirement-related decision-making. Pension communication policies accounting for individuals' lack of involvement and expertise are a second way to achieve the goal. The third tool is choice architecture: organizing the choice process: the setup, presentation, and timing of the choices that individuals need to make. Several MOPACT studies investigated the roles of financial literacy, pension communication, and choice architecture in improving individuals' pension and retirement decisions.

Optimal Contribution, Drawdown, Investment, and Insurance

Introduction: Unbundling

In this section, we will summarize the existing academic literature and the current pension products and legislation in terms of contribution, drawdown, investment, and insurance strategies and outline the new findings in the MOPACT project. A first and perhaps the most important insight from the MOPACT project is that these strategies can be unbundled. Often the policy discussion has been on all these decision variables mixed together, because one focused on specific pension systems. One specific example is a pure DB system with mandated participation with prescribed contribution rates that provides income guarantees and where all biometric risks are shared. Another would be a DC system with freedom of choice on contribution and drawdown rates, no sharing of longevity risk, and lots of uncertainty with respect to future income for the participants. Bovenberg and Nijman (2016) propose to unbundle these choices using the concept of Personal Pension with Risk-sharing (PPR).

Just as in an individual DC scheme, an individual features a personal claim on financial assets in a PPR. These assets are the property of the individual; a PPR is a personal account shielded from the investment and (dis) saving decisions of others. A PPR is a *personal* pension. Yet, an individual cannot freely dispose of the funds because the personal assets are earmarked for retirement income and therefore cannot be claimed for consumption before retirement. Accordingly, a PPR is a personal *pension*. As a third element, a PPR may include insurance of (micro and macro) longevity risk and other contracts pooling idiosyncratic risks and/or trading systematic non-financial risks. A PPR is a personal pension *with risk-sharing*.

During the accumulation phase, contributions flow into the PPR. Retirement income is drawn from the PPR during the decumulation phase. In each period, three types of returns (financial, biometric, and non-market returns) add to the account. The financial assets in the personal account generate the financial returns, just as in personal DC plans. The insurance elements in the PPR yield biometric returns. These returns are conditional on idiosyncratic risks faced by the policyholder. With longevity insurance, for example, an owner who survives enjoys a positive biometric return but the assets accrue to the insurer if the owner passes away. Returns on non-market assets can be generated by risk-sharing agreements within the pool written on systematic risks that are not traded on financial markets (such as (wage) inflation or a change in life expectancy). Such insurance elements can but do not have to be incorporated. They can also be included at specific ages, for example, longevity insurance can be provided only at older ages.

The PPR concept is flexible enough to allow for alternative mechanisms for absorbing mismatch throughout the payout period. One option is to absorb shocks instantaneously in accordance with the Merton–Samuelson model. An alternative option is to absorb shocks gradually. Likewise, the PPR concept can either be implemented including longevity risk insurance (as in annuities) or without and the income generated by the PPR can contain guarantees if these are preferred.

PPR solutions can be tailor-made in that the (dis)saving, investment, and insurance strategies are based on personal preferences and characteristics. Whereas traditional DB pensions and some popular advice on DC pensions are largely 'one size fits all', van Ewijk et al. (2015) show that substantial welfare gains can be obtained by tailoring these strategies to the age of the participant (as in lifecycle investing) but also to existing pension entitlements and, for example, to whether one is houseowner and or prepays a mortgage.

An additional feature of a PPR is that the product is flexible and does not rely on long-term contracts with pension providers. The property rights are with the individual as is the case with mutual funds. The PPR concept underlies the recent (September 2016) regulatory adjustment to the decumulation phase of DC pensions in the Netherlands, where now flexible variable annuities can be offered. The concept is also at the heart of the ongoing discussions of reform of the Dutch DB pensions. A more detailed discussion of the PPR concept is provided in Bovenberg and Nijman (2016). A more technical discussion is available in van Bilsen and Bovenberg (2016).

In the remainder of this section, we consider the decisions on (dis)saving, investment, and risk-sharing one by one. In the next section, we start with the savings decision and investment during the accumulation phase and summarize the takeaway messages from the MOPACT project on comparing the regulations in five European countries and the analysis of the proposed PEPP regulation. Then we turn to the decumulation phase and consider optimal design and regulation of risk-taking drawdown products. In particular we emphasize how regulation on the maximum Assumed Interest Rate (AIR) can be used to materialize the policy goal of affordable stable pensions either for a fixed number of years (i.e. without longevity insurance) or lifelong (with longevity insurance). We then extend this analysis to the smoothing of shocks in pension capital which is optimal in case of preferences involving habit formation. Finally, we consider further extensions of this basic framework that have also been investigated in the MOPACT project, in particular the preference for guaranteed income streams.

Savings and Investment Decisions in the Accumulation Phase

The savings and investment decisions in the accumulation phase of funded pension schemes have been extensively analyzed in the academic literature. The work horse model is the so-called Merton–Samuelson model which dates back to Merton (1969) and Samuelson (1969). The model assumes risk-free human capital, that is, number of hours and years worked are known in advance and the wage level is fixed. The date of death is deterministic. The model has only one risk factor, equity risk. The financial market satisfies the assumptions made by Black and Scholes in deriving their option pricing formula, that is, interest rates are fixed and equity (log) returns are i.i.d. normal. Agents are characterized by time separable power utility functions and can choose how much to save and how much of their pension wealth to invest in the risky asset. This standard model implies that it is optimal to invest a lot (even more than 100% of your pension wealth) when young and gradually reduce the equity exposure when one approaches the retirement age. The model has been extended in many different directions, for example, using other preference assumptions, other assumptions on financial markets, and other assumptions on the labor market behavior (see Bovenberg et al. 2007, for an extensive survey). Nevertheless, the model underlies most of the popular savings and investment advice. In particular the dominance of target date funds and lifecycle investments seems to be largely based on the recommendations of the Merton–Samuelson model.

In one of the MOPACT papers (Nijman et al. 2015), we have compared the recommendations on the design for PEPPs in the EIOPA consultation paper with the literature. The policy documents emphasize the relevance of lifecycle investing but provide little guidance on how much to invest in equities at what age. A second key recommendation is to restrict encashing pension wealth before retirement. The MOPACT paper referred to above has identified that here as well current legislation differs widely between European countries. In Finland, for example, early withdrawal of pension wealth is allowed in case of long-term unemployment or divorce. In Hungary, it is costly to take up pension wealth before retirement and there is a minimum accumulation period for that of ten years. In Estonia and the Netherlands, it is not possible to access pension capital before retirement. The recommendation to restrict encashing pension wealth before retirement is in line with the academic literature. Beshears et al. (2015) provide a comparison of the current legislation on early withdrawals in a number of other countries as well as for arguments in favor or against provision of opportunities for early withdrawal.

Dissaving, Investment, and Insurance Decisions in the Decumulation Phase

As stated before, the diversity between countries in regulation and recommendations on (dis)saving, investment, and insurance decisions in the decumulation phase is much larger than in the accumulation phase.[1] Where some countries allow full freedom whether or not to assure life-long pension income by insuring the (micro) longevity risk of getting very old, others mandate specific forms of annuity contracts. Likewise, guaranteed income streams are the dominant pension solutions in some countries, but more and more variable annuities are promoted as a preferable alternative.

The standard Merton–Samuelson model referred to above suggests a constant equity exposure during retirement, the level of which depends on parameters such as the risk preference and the Sharpe ratio. The model assumes a deterministic date of death and is therefore not suited to address the question how to allocate biometric returns in the PPR, that is, whether or not to insure longevity risk. There is an extensive academic literature that addresses this question though. Under standard assumptions, a constant equity exposure will still be optimal in a Merton–Samuelson type model. Moreover, this literature finds substantial welfare gains of insurance of longevity risk, unless, for example, one has strong bequest motives, cost charges for annuitized products are very high, or only a limited range of financial products is available with longevity insurance. Peijnenburg et al. (2017) provide recent additions to the literature as well as an up-to-date literature review.

A natural policy goal is to have adequate and not too costly stable pension income (see also EIOPA 2015). Irrespective of whether one prefers to insure longevity risk (countries like Denmark and the Netherlands) or aims at pension income for a fixed number of years only where remaining pension wealth would be transferred to the heirs, the drawdown decision and investment can be structured and regulated in a very similar way. As discussed in Bovenberg and Nijman (2016) and in Balter and Werker (2016), a key parameter here is the AIR. The AIR is used to determine the current pension income to be drawdown from the current pension capital. The name AIR can be explained by the property that if in all subsequent years the actual return on risk investment equals the AIR, the pension income will be constant in all these years (and the consumption to wealth ratio with the same AIR will be higher if longevity risk is insured).

Balter and Werker (2016) illustrate the impact of choice and regulation of the AIR using a simple numerical example where the aim is to have adequate pension income for a 20-year period. They assume a constant equity exposure of 25% during retirement (consistent with the extended Merton–Samuelson model), an equity premium of 4%, an interest rate of 1%, and a volatility of equity returns of 20%. Figure 4.1 illustrates the implications of choosing the AIR equal to the expected return on the portfolio with 25% equities, that is, AIR = 1% + 0.25 * 4% = 2%.

Figure 4.1 clearly shows the important feature that the expected pension income is constant if the AIR equals the expected portfolio return, which is one aspect of the policy goal to have adequate, not overly costly and stable pension income. Until September 2016, in the Netherlands DC pensions had to be decumulated as nominal guarantees (see the line 'fixed income' in Fig. 4.1). As of then, the pension capital can also be

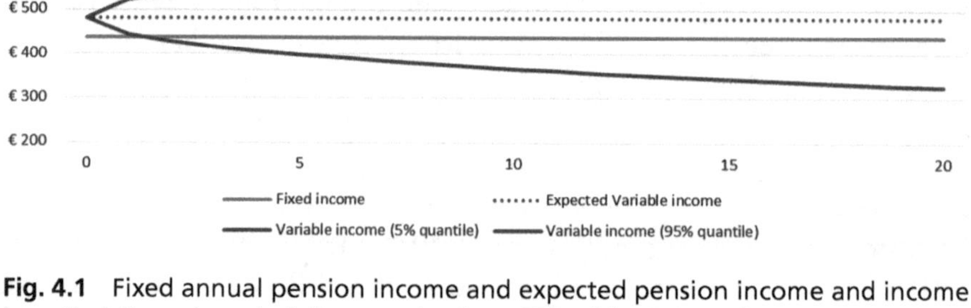

Fig. 4.1 Fixed annual pension income and expected pension income and income in optimistic and pessimistic scenarios for a PPR with 25% equity exposure where the AIR equals the expected return on the portfolio

decumulated as the product structures and the AIR reflected in Fig. 4.1. Note also that this figure contains confidence bands for the pension income in future years, which can be of great help to explain the advantages and disadvantages of risky investments to plan participants. Soon all Dutch pension providers will have to include these under the heading of 'optimistic scenario' and 'pessimistic scenario' in their communications with participants and clients.

By choosing the AIR, plan participants (or the trustees of a plan, or the regulator) can set the level of the first payment in the drawdown period as well as the expected future path of these payments. To illustrate this, Fig. 4.2 presents the same information as before using the same assumptions but now with AIR = 1%.

In this case, the AIR is less than the expected return on the portfolio which implies an increasing expected future income stream in case of constant equity exposure. Likewise, an AIR that exceeds the expected return will imply a decreasing expected future income stream. To protect the lifelong stable character of the income stream, the AIR that clients and providers are allowed to select can be capped. The Dutch regulator has imposed a cap on the AIR equal to the expected return on the portfolio.[2] Note that the Dutch legislation contains incentives for risk-taking

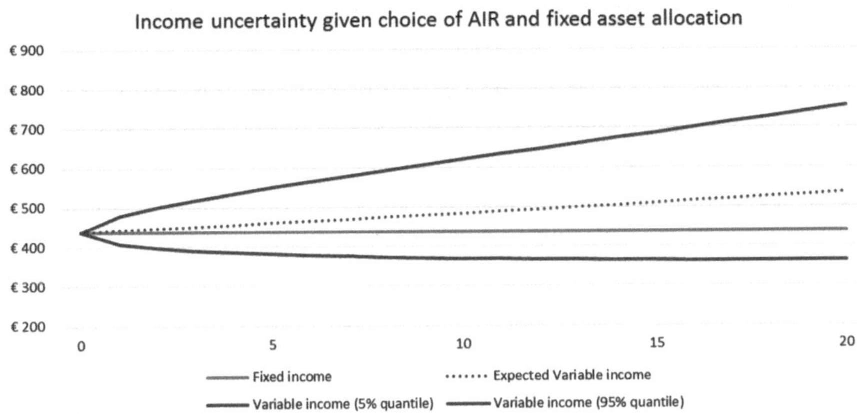

Fig. 4.2 Fixed annual pension income and expected pension income and income in optimistic and pessimistic scenarios for a PPR with 25% equity exposure where the AIR equals the risk-free rate

for myopic agents that are primarily concerned with the income level in year one. For this reason, and to discourage excessive risk-taking, the regulator has also capped the AIR that can be used at the expected return of a portfolio with 35% exposure to equities. More than 35% exposure to equities in the decumulation phase is allowed (subject to prudent person rules) but is capped in setting the AIR.

A simple rule of thumb on the impact of the fraction invested in equities on the first pension payment if the maximum AIR is used can be derived using duration analysis. The duration of pension payments over a 20-year period is approximately 10.[3] Whereas the payment in case of guaranteed pension income is determined using a risk-free discount rate, the stable payment level in case of risk-taking in the decumulation phase is determined using the expected portfolio return as the discount rate. The total impact on the payment level equals the product of the duration and the difference in the discount rate. In the example discussed here, we find an impact of 10 * 1% = 10% in case of 25% exposure to equities, which is consistent with Fig. 4.1.

Dissaving, Investment, and Insurance Decisions in the Decumulation Phase in Case of Habit Formation

In several papers that are directly related to the papers in the MOPACT project, we have extended the results in the previous section to more general assumptions on preferences (see e.g. van Bilsen et al. 2016). Here we consider the case of preferences with habit formation and later the case of a preference for income guarantees. As in the previous section, one can implement these strategies either with or without longevity insurance.

The solution of the Merton–Samuelson model outlined earlier yields rather volatile pension income if substantial investment risk is taking in the decumulation phase. In order to remedy this, one can allow for smoothing of shocks.[4] Shocks in pension capital are gradually transferred over an N-year period to shocks in pension income, enabling the combination of low volatility pension income with substantial investment risk. Consider, for example, a portfolio return of −10%. Instead of reducing the pension payment immediately by 10%, it is reduced by 1% (assuming N = 10), again by 1% in the next year, and so on. Clearly, this implies that the projected payments in year n of the retirement phase ($n > N$) have to be cut by more than 10% to fulfill the budget constraint. Bonenkamp et al. (2016) and Balter and Werker (2016) consider a drawdown and investment strategy where shocks are smoothed over N years, the year-to-year volatility in the pension payments is constant and the expected pension payments are constant. In order to satisfy the budget constraints, the reduction in remaining life expectancy implies lifecycle investing if the goal is to have constant year-to-year volatility in pension income. Note that these lifecycle investments in de decumulation phase are generated by the smoothing mechanism and a preference for constant year-to-year volatility, unlike the more common lifecycle strategies in the accumulation phase which stem from the assumption of risk-free human capital.

Figure 4.3 compares the income uncertainty of strategies with and without smoothing of shocks using the mechanism and parameter

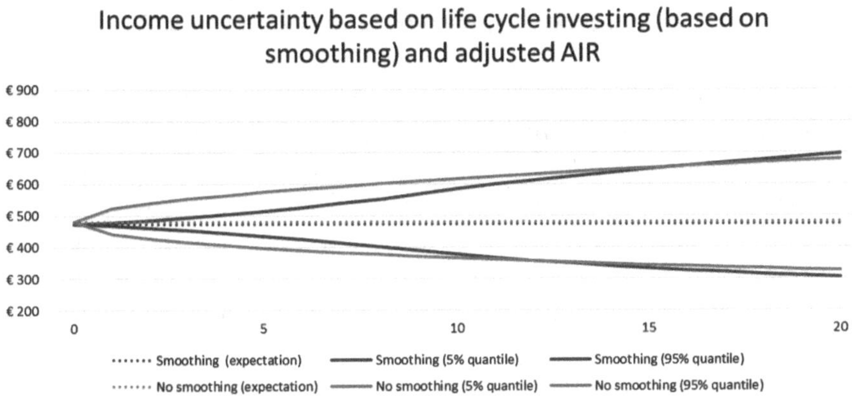

Fig. 4.3 Expected pension income and income in optimistic and pessimistic scenarios for PPRs with and without smoothing during ten years. The strategy without smoothing uses a 25% equity exposure and AIR equals the expected return on the portfolio. The strategy with smoothing uses the lifecycle strategy as described and an adjusted AIR

assumptions referred to above. Smoothing reduces short-term risk at the expense of long-term risk. The AIR here implies constant expected future income and, as before, equals the expected return on the investment strategy. Because of the lifecycle investments, this AIR (or discount rate) now depends not only on the level of risk-taking but also on the horizon. For short horizons, less investment risk is taken and therefore a lower risk premium is to be incorporated in the discount rate. See Balter and Werker (2016) and Nijman et al. (2012) for more details.

Dissaving, Investment, and Insurance Decisions in the Decumulation Phase in Case of Income Guarantees

In contrast to the recommendations of the standard Merton–Samuelson model, pension products that offer income guarantees in the decumulation phase have been dominant in many countries. Van Bilsen et al. (2014) describe preference assumptions based on prospect theory that rationalize a preference for guarantees. They pragmatically consider the availability of income guarantees and analyze how a preference for guarantees affects the dissaving, investment, and insurance decisions in the decumulation phase.

Income guarantees are costly in a low interest rate environment because budget constraints imply that they have to come with risk-free investing. As argued before, the remaining life expectancy at retirement is 20 years, the duration of the income stream is approximately 10. At an interest rate of say 0.5% (January 2017) a given level of pension capital will therefore buy some 35% less guaranteed income compared to an interest rate of 4% (the actual interest rate a few years ago). As discussed above, the initial income in case of risk-taking in the decumulation phase will typically be higher, because of a very similar reason, the use of a higher AIR (or discount rate).

Of course, and as illustrated in Figs. 4.1, 4.2 and 4.3, risky investing in the decumulation phase comes with the probability of lower pension income than would have been generated by guarantees. Agents with different preference functions or different degrees of risk-aversion will prefer different drawdown and investment strategies. The new Dutch DC regulation offers agents the choice for either the traditional income guarantees (the default) or the risk-taking PPR structures outlined before. Note that one can also envisage products that combine a guaranteed income floor with risky investing during retirement (see e.g. Balter and Werker 2016; van Bilsen et al. 2013).

Another very important aspect of the drawdown and investment strategies in the decumulation phase is how these are aligned with the investment strategy in the last years of the accumulation period. Agents taking substantial equity risk shortly before retirement can be confronted with substantial unpleasant surprises when they retire, because the actual pension income can differ substantially from what was projected only shortly before retirement. This risk is often referred to as conversion risk. The conversion risk on equities can be reduced using a lifecycle strategy that sets the equity exposure in the last years before retirement equal to the equity exposure after retirement. Conversion risk is even more important if interest rates are stochastic. As outlined above, a drop in interest rates can have a large impact on the income level in case of income guarantees and the same is true in case of risky investing, as is evident from the expressions for the optimal AIR. Hedging interest rate risk in the (final years of) the accumulation phase remedies this problem. The Dutch prudential supervisor (AFM) imposes that both equity risk and interest rates risk in the conversion from accumulation to decumulation have to be managed.[5]

One of the MOPACT projects, Nijman et al. (2015), analyzed EIOPA's consultancy paper on PEPPs. One of the main focus points was the lack of attention for how the investment strategy in the accumulation phase is to be tuned to the requirements of the decumulation phase. Stated differently, this chapter argued that the consultation paper treats pension products as if they would be asset only,[6] while in fact the goal of stable income requires them to be ALM products where the investment strategy is to be aligned with the liabilities. The pre-occupation in the PEPPs with the accumulation phase is likely to be caused by the very different regulatory requirements and preferences on drawdown products in different European contracts. The EU goal of stable pension income is reflected in some countries by regulatory incentives to have drawdown products that generate stable income over a fixed number of years. Other countries mandate lifelong income streams. Moreover, many countries allow full withdrawal of pension capital at retirement. Different strategies in the decumulation phase will require different strategies in the (last years of) the accumulation phase to limit conversion risks. One of the attractive features of the unbundling of strategies offered by PPRs though is that the same basic framework can be used to regulate conversion risk for many different choices of the decumulation strategy, using different parameter assumptions reflecting the number of years for which the stable income is to be generated, whether income guarantees are offered and at what level and which biometric risks are shared.[7]

Individual Pension Decision-Making

The recent literature in economic psychology and behavioral economics demonstrates convincingly that individuals do not always make optimal decisions, that is, decisions that are in their best interest. It has been shown that the quality of their decisions is influenced by (lack of) financial literacy, (lack of) interest and involvement, communication and framing of the decision task, and the timing and context of the decision. In this section, we discuss these aspects in relation to decisions relating to pensions. The pension wealth decumulation choices addressed in the previous section are prominent examples of such decisions, but the findings in this section also apply to many other pension-related decisions.

Financial Literacy and Pension Literacy

Financial Literacy

Recent studies covering many different countries have demonstrated that many people are unable to answer simple questions on compound interest rates, inflation, or risk diversification (see e.g. Lusardi and Mitchell

2011). In other words, many people lack basic financial literacy. Moreover, substantial differences in financial literacy and illiteracy exist across countries but also across socioeconomic groups, generally indicating, for example, low financial literacy among women, among individuals with low education level, and among the oldest and youngest age groups.

The data show that financial literacy is positively associated with retirement planning, that is, with thinking about the financial resources needed for retirement. This does not necessarily imply that improving financial literacy helps to improve pension decision-making, since there may well be confounding factors or reverse causality. Several different identification strategies have led to the common conclusion that, indeed, there is a positive causal effect of financial literacy on pension planning (e.g. Alessie et al. 2011). Moreover, existing studies show that more retirement planning leads to more pension wealth accumulation. This link is weaker, though, since the behavioral economics literature has clearly demonstrated that there is a large difference between intentions and actual behavior. Financial literacy also positively affects other financial decisions that can improve retirement resources, such as stock market participation (van Rooij et al. 2011) or on-the-job training (Clark et al. 2012).

The research on the effectiveness of interventions aimed at improving financial literacy has not led to a clear conclusion yet. The idea of mandatory financial education has met some support but also strong opposition, due to the costs and the limited improvement that can be expected. A meta-study of more than 200 articles even suggests that interventions aimed at improving financial literacy have hardly led to any improvement in financial decisions (Fernandes et al. 2014).

A promising approach to get insight in the role of financial literacy seems to be to model financial literacy as one of the endogenous variables in a lifecycle model, where individuals make the trade-off between the costs and the expected future benefits of investing resources (time and money) into financial literacy. MOPACT researchers Jappelli and Padula (Jappelli and Padula 2013) develop an intertemporal model explaining consumption, wealth accumulation, and investment in financial literacy where the gains of financial literacy take the form of higher returns on financial wealth. Solving their model, they find a positive relation between financial literacy and wealth over the lifecycle. Moreover, the incentive to invest in financial literacy is stronger in countries with lower automatic pension savings. Using micro-data from the Survey of Health, Ageing and Retirement in Europe (SHARE) supplemented with other data sources, they find strong empirical support for the predictions of their model. More sophisticated models in the same spirit are introduced in Lusardi et al. (2017). Their model explains why in many pension systems, low-income earners, for whom the automatic state pension provides most of their pension income, have less to gain from investing in financial literacy. As a consequence, even if preferences and investment costs for financial literacy do not vary with income, higher-income groups invest more. In the long run, this also implies they accumulate more wealth. Lusardi et al. (2017) also emphasize the importance of the timing of the investment in financial literacy, implying that financial education is effective only if offered shortly before periods where financial decisions involving financial literacy really matter. Effective financial education programs must incorporate follow-up such that the initial investment is not lost after a few years. For some groups of individuals, an initial endowment of financial literacy can make further investments worthwhile, implying a multiplier effect that long-run returns to financial education can exceed the short-term gain by far. Their model simulations (based upon plausible calibrated parameter values) suggest that, in the US pension system, financial literacy programs offered at the right time (optimally at round age 40) could increase retirement savings substantially. These models clearly demonstrate that the relation between financial education and financial literacy is far from straightforward and, depends on timing, and differs across groups.

Pension Literacy

To make pension-related decisions leading to an adequate pension, individuals need specific knowledge concerning their country's pension system and their own entitlements and choice options. As expected, the literature has demonstrated that there is a positive causal effect of pension knowledge on pension savings; see, for example, Landerretche and Martinez (2013) for evidence in Chile, where employees have mandatory retirement savings, but can choose where to invest. In line with the theory, the literature shows that pension knowledge is larger in countries with an individual (rather than collective) pension system, where individuals themselves have larger responsibility for their pension savings (Towers Watson 2012).

The existing studies show that pension knowledge is limited and interest in pension issues ('involvement') is low among many socioeconomic groups (particularly the younger age groups) in all countries. Many individuals cannot forecast their retirement income and do not know what actions could raise their old-age standard of living. This is also the case in countries like Sweden or the Netherlands, where individuals get annual updates about accumulated pension rights. Lack of interest and involvement is one possible explanation for this. Another possible reason is that pensions and pension rules are extremely complex and difficult to understand, while pension communication focuses on fulfilling legal obligations instead of helping individuals to make adequate decisions. Whether pension communication induces individuals to get involved and change their behavior strongly depends upon the presentation and wording of the message and not only on its objective content. The MOPACT study by Boggio et al. (2014) suggests that the sense of familiarity induced by the framing of the message may have an important effect on the choice that an individual makes. They demonstrate that in the financial world, language is dominated by metaphors that are more familiar to men than to women (terms taken from war, sports or games, like 'beat the market') rather than female and gender-neutral metaphors (referring to health or farming, like 'healthy investment' or 'portfolio yield'). They conclude that this can explain the gender gap between financial market participation of men and women.

Choice Architecture

Numerous experiments have shown convincingly that individuals are very sensitive to what for a purely rational individual should be irrelevant aspects of the way choices are organized and presented. This applies not only to financially illiterate people but also, albeit to a lesser extent, to the financially literate. It certainly applies to choices involving difficult intertemporal tradeoffs and uncertainty such as many pension-related decisions. It implies that financial literacy and pension knowledge are not sufficient to achieve the goal of adequate retirement preparation.

Kahneman (2011) presents a conceptual framework where individuals can make decisions in ways that vary from consciously and rationally ('thinking slow' or 'system 2 thinking' in Kahneman's terms) to unconsciously and intuitively, based upon rules of thumb or other choice contextual aspects ('system 1 thinking'). This builds on the famous Elaboration Likelihood Model of Petty and Cacioppo (1986) in economic psychology and marketing, where individuals who are not involved or lack the necessary ability make their decisions following a 'peripheral route' (system 1) while optimizing rational behavior follows the 'central route' (system 2). The way in which the choice is organized and presented is called choice architecture. Choice architecture can be used to bring individuals closer to the central route, increasing the likelihood that they make a decision that is in their own best interest. Choice architecture also determines the contextual aspects of the choice, and therefore affects the decisions of those who have taken the peripheral route. It can therefore nudge people toward making a certain choice. For example, it is well established that many young people lack the motivation to think about pensions, since their old age is beyond their time horizon. Their future-selves can be made more salient by providing them with an artificial picture of themselves (an 'avatar') at old age in a certain mood or in certain circumstances. Hershfield et al. (2011) have shown experimentally that this helps to increase their retirement savings. It is an example of how the presentation and context of the choice can stimulate individuals' involvement, bring them closer to the central route, and improve the quality of the decision. Prast and van Soest (2016, Box 2) give more examples in the domain of saving for retirement. They emphasize the importance of defaults as one of the aspects of choice architecture. What is the passive choice if the individual does not undertake any actions (and does not make an active choice)? But this is not the only tool that can nudge individuals toward making better decisions. Some other relevant tools are: splitting up complex choices into several not so complex steps, presenting choices in such a way that certain aspects of a choice become more salient, and eliminating the passive choice (and thus forcing individuals to make an active choice). These tools do not affect the actual choice set, only the choice process, presentation, and context. This is what choice architecture is about; it is a form of 'soft paternalism' that does not impose any restrictions on what people can choose.

In addition, the choice set itself can be adapted to help individuals to make better choices. This can work in two ways. First, inferior products that cannot be optimal for any individual should not be offered. Second, it may be useful to offer specific products that are designed to make use of non-rational aspects of choice behavior. An example in the United States concerning voluntary pension wealth accumulation is 'Save more tomorrow' (Thaler and Benartzi 2004), a pension savings product inspired by the tendency to procrastinate and the need of a commitment device: individuals do not commit to start saving immediately, but with some delay, and the amount of saving is linked to their increase in earnings. Since this does not affect current expenditure, many more individuals are willing to commit to such a savings plan than to a plan that requires immediate saving.

Implications for Policy and Practice

The MOPACT project has delivered at least partial answers with a lot of relevance for policy and practice on the two key challenges that were outlined at the start of this chapter. Risk-taking in the decumulation phase of DC pension products can very well be motivated, but restrictions on the drawdown strategy, for example through caps on the AIR, seem desirable to achieve the (personal and) policy goal of stable income streams for retirees. The upcoming regulations for PEPPs are to be adapted to these findings. Doing so seems relatively straightforward using the concept of PPR.

There is ample evidence in the literature that fine-tuning the design of pension products to individual characteristics can yield substantial welfare gains. This is also the case for the decumulation phase of DC pensions. A much more open question though is how to structure the decision processes that are to deliver this fine-tuning. Freedom of choice on many design issues is one of the policy options but also has a disadvantage if lack of involvement and low financial and pension literacy prevent individuals from making the decisions that are in their own best interest. In the MOPACT project, we therefore also analyzed the alternative route of using trustees of pension providers and duty of care legislation to come to a choice architecture that will lead to better decisions.

Policymakers have been concerned with pension communication since a long time, and many countries have laws that require pension providers to inform their customers in a correct and transparent way. Until recently, pension providers often saw this as a legal obligation. This trend seems to be changing in some countries. First, the parties involved now often realize that informing individuals is not the final goal. Information and communication should help individuals to make better decisions. Second, the notion that the *homo economicus* does not exist in reality is gaining ground. Policymakers and sector practitioners now have a better understanding of what is adequate communication and which behavioral biases can arise in the pension decisions of individuals. Communication strategies, for example, focus more on getting people involved and on transparency for individuals with low pension literacy. The United Kingdom has started using choice architecture by changing standard defaults for voluntary pension contributions.

Considering the goal of improving individuals' pension decisions, either in the accumulation or in the decumulation phase, we expect much more from choice architecture than from the other tools. Financial literacy positively influences involvement with pensions and the quality of pension-related decisions, but improving financial literacy through education is less straightforward. Financial education programs may not always be effective, and some studies consider them as an expensive policy with uncertain payoff. Improving financial literacy and pension knowledge is certainly not sufficient. Regulations need to prevent the possibility of choosing inferior pension products or to make choices that increase the risk of old-age poverty. This justifies, for example, mandatory minimum pension contributions and partner pensions, or an upper limit on the share of risky pension investments in the decumulation phase.

Offering well-designed saving products such as 'Save more tomorrow' that account for individuals' behavioral biases seems quite promising well. In addition, choice architecture must guarantee that individuals are nudged to making choices that are close to optimal, by thinking carefully about how to present and organize the choice process.

For the practical implementation, an important issue is who does what. In the accumulation phase, several parties have their own 'duty of care' when it comes to helping individuals to adequately prepare for retirement: the government and the regulator, social partners, employers, pension funds, insurance companies, financial advisors, and the individuals themselves. The way this is organized varies across countries and pension systems, and changes in the system may require rethinking the roles of the parties involved.

Specifically, this also applies to the unbundling of functions in a PPR. The transparent market-based valuation of a PPR implies that decentralized parties can make choices without imposing externalities on others. Hence, a PPR allows responsibilities to be delegated to decentralized parties, such as mutual insurers, social partners, employers, and individuals. At the same time, consistent with the subsidiarity principle, different parties can select and administer the various unbundled functions depending on the institutional structure and history of a particular country. In fact, various parties (employers, unions, organizations of self-employed, civic society, pension funds, commercial insurers, and government) can play a role in deciding on a particular function—with a different mix of responsibilities in each country, depending on social preferences and institutions.

We can illustrate the flexibility to allocate responsibilities to various parties with the savings function. The government may set limits for tax deductible pension contributions, social partners may negotiate the choice architecture (including default contributions and employer incentives to contribute), and individual workers may opt out of (part of) these contributions. Alternatively, the government can set a minimum contribution rate (as in Australia) or it can force employers to offer a minimum default contribution rate to their workers (as in the United Kingdom). The transparent link between contributions and accrued capital in a PPR makes voluntary contributions more attractive for the self-employed, who at present often accrue substantially fewer pension rights than employees do.

PPR structures can accommodate different parties arranging risk-sharing. In particular, insurance pools may be formed by commercial insurers through competition on the market for individual longevity insurance. Alternatively, social partners or professional groups can form these pools. The government (e.g. regional governments) may also play a role by pooling longevity risks in the decumulation phase, while the private sector manages the financial assets in PPR. The public sector then essentially acts as a mutual insurer, while individuals or collectives (e.g. organized by social partners) can select asset managers and the risk profile of investments. The premium pension in Sweden comes close to this construction.

Further Research Priorities

The most pressing priorities for further research have to do with realistic fine-tuning of the (dis)saving, investment, and insurance decisions of individuals. Agents as well as pension providers will have heterogeneous information on individual life expectancy, on whether or not agents have a safe job until retirement, on current wealth and housing value, and so on. Financial advice as well as regulatory practice will also have to be adjusted to the new realities of big data where more and more information on individuals is somehow available. The MOPACT study of Maatanen et al. (2014) showed that changes in life expectancy are dealt with in very different ways in different countries. Little attention, however, has been paid to the implications of biometric risks (risk of getting old, risking of losing your partner), changes in overall life expectancy, and improving opportunities to predict one's own life expectancy for an adequate design of pension products. This clearly requires additional research. Another clear challenge for research remains how to structure the choice architecture in an optimal way to stimulate individuals to make active and rational decisions in their own best interest. This requires an interdisciplinary effort, where expertise of psychologists, economists, lawyers, and others can fruitfully be combined to make progress.

Conclusions

In this MOPACT project, researchers from a large number of European countries have actively worked together. Apart from the important networking element to meet researchers working on topics related to your own expertise, the project has produced more in debt knowledge and understanding of the many important differences in institutional arrangements in different European countries. Research-wise the project has generated a large number of academic papers that add to the understanding of the key challenges for the project. Some of these results have been summarized in this chapter. As always in research, new questions have appeared, and not all questions could be answered. Nevertheless, we are confident to have shifted the frontier of knowledge forward in ways that can be beneficial for pensioners in the EU and worldwide.

Notes

1. As argued in Nijman et al. (2015), it is, moreover, important to align contribution and investment strategies in the accumulation phase to the income goals in the decumulation phase, which is identified as one of the most important shortcomings in the EIOPA proposals for PEPPs. This issue will be discussed in the section on dissaving, investment, and insurance decisions, which also considers interest rate risk.
2. Balter and Werker (2016) show that in the context of the Merton–Samuelson model the optimal level for the AIR is slightly less than the expected portfolio return.
3. The duration of a cash flow is a measure for the sensitivity of the economic value of the cash flow to the discount rate that is assumed. The duration can be approximated by the average number of periods until each of the payments in the cash flow is due.
4. This option has been introduced in the new Dutch DC regulation referred to before. Note that in Dutch DB (or CDC) funds shocks are smoothed in a very similar manner.
5. The current rules relate to income guarantees. The rules will probably be adapted to the choice options offered in the newest DC regulation.

6. This is also reflected in the proposal to use the Key Information Document developed by the EIOPA for communication purposes. This document lacks information on long-term risk factors such as inflation risk and (macro)longevity risk and lacks information of biometric micro-risks.
7. Note though that this requires assumptions before the retirement date on the decumulation strategies that will only be set at the retirement date and can potentially be adjusted during retirement.

References

Alessie, R. M., van Rooij, M., & Lusardi, A. (2011). Financial Literacy, Retirement Preparation and Pension Expectations in the Netherlands. *Journal of Pension Economics and Finance, 10*(4), 527–545.

Balter, A., & Werker, B. (2016). Variable Annuities in the Dutch Pension System, Manuscript.

Beshears, J., Choi, J. J., Hurwitz, D., Laibson, D., & Madrian, B. C. (2015). *Liquidity in Retirement Savings Systems: An International Comparison* (NBER Working Paper, 21168).

Boggio, C., Fornero, E., Prast, H., & Sanders, J. (2014). *Seven Ways to Knit Your Portfolio: Is Investor Communication Neutral?* CeRP Working Paper, CeRP, Turin, 140(14).

Bonekamp, J., Bovenberg, A. L., Nijman, T., & Werker, B. (2016). *Welke vaste dalingen en welk beleggingsbeleid passen bij gewenste uitkeringsprofielen in verbeterde premieregelingen* (Netspar Occasional Paper (Dutch)).

Bovenberg, A. L., & Nijman, T. (2016). *Personal Pensions with Risk Sharing" Forthcoming in Journal of Pension Economics and Finance* (Revised Version of Netspar Discussion paper).

Bovenberg, A. L., Nijman, T., Teulings, C., & Koijen, R. (2007). *Saving and Investment Over the Lifecycle: The Role of Individual and Collective Pension Funds* (Netspar Panel Paper nr. 1).

Clark, R., Morrill, M. S., & Allen, S. G. (2012). The Role of Financial Literacy in Determining Retirement Plans. *Economic Inquiry, 50*, 851–866.

EIOPA. (2015). *Consultation Paper on the Creation of a Standardized Pan-European Personal Pension Product (PEPP)*. Frankfurt: EIOPA.

Fernandes, D., Lynch, J., & Netemeyer, R. (2014). Financial Literacy, Financial Education and Downstream Financial Behaviors. *Management Science, 60*(8), 1861–1883.

Hershfield, H., Goldstein, D., Sharpe, W., Fox, J., Yeykelis, L., Carstensen, L., & Bailenson, J. (2011). Increasing Saving Behavior Through Age-Progressed Renderings of the Future Self. *Journal of Marketing Research, 48*, S23–S37.

Jappelli, T., & Padula, M. (2013). Investment in Financial Literacy and Saving Decisions. *Journal of Banking and Finance, 37*, 2779–2792.

Kahneman, D. (2011). *Thinking Fast and Slow*. New York: Farrar, Straus and Giroux.

Landerretche, O., & Martínez, C. (2013). Voluntary Savings, Financial Behavior and Pension Finance Literacy: Evidence from Chile. *Journal of Pension Economics and Finance, 12*, 251–297.

Lusardi, A., & Mitchell, O. (2011). Financial Literacy Around the World: An Overview. *Journal of Pension Economics and Finance, 10*, 497–508.

Lusardi, A., Michaud, P.-C., & Mitchell, O. S. (April 2017). Optimal Financial Knowledge and Wealth Inequality. *Journal of Political Economy, 125*(2), 431–477.

Määttänen, N., Võrk, M., Piirits, R., Gal, E., Jarocińska, A., Ruzik, A., & Nijman, T. (2014). *The Impact of Living and Working Longer on Pension Income in Five European Countries: Estonia, Finland, Hungary, the Netherlands and Poland* (Netspar Discussion Paper, 36).

Merton, R. C. (1969). Life Time Portfolio Selection Under Uncertainty: The Continuous Time Case. *Review of Economics and Statistics, 51*, 247–257.

Nijman, T., Werker, B., & Bovenberg, A. (2012). *Voorwaardelijke pensioenafspraken: Over waarderen, beschermen, communiceren en beleggen* (Netspar Occasional Paper).

Nijman, T., Määttänen, N., Vork, A., & Gal, R. (2015). *Analysis of the Standardized Pan European Personal Pension (PEPP) Product and Its Impact in Four European Countries: The Netherlands, Estonia, Finland and Hungary*. (Netspar Discusssion Paper, 11, 64).

Peijnenburg, J. M. J., Nijman, T. E., & Werker, B. J. M. (2017). Health Cost Risk: A Potential Solution to the Annuity Puzzle. *The Economic Journal, 127*(603), 1598–1625.

Petty, R., & Cacioppo, J. (1986). The Elaboration Likelihood Model of Persuasion. *Advances in Experimental Social Psychology, 19*, 123–195.

Prast, H., & van Soest, A. (2016). Financial Literacy and Preparation for Retirement. *Intereconomics, 51*(3), 113–118.

Samuelson, P. A. (1969). Life Time Portfolio Selection by Dynamic Dynamic Stochastic Programming. *Review of Economics and Statistics, 51*, 239–246.

Thaler, R., & Benartzi, S. (2004). Save More Tomorrow™: Using Behavioral Economics to Increase Employee Saving. *Journal of Political Economy, 112*(S1), S164–S187.

Towers Watson. (2012). *Global Workforce Study 2012 Nederland: Hoe denkt werkend Nederland.* Retrieved from http://www.towerswatson.com/nl-NL/Insights/IC-Types/Survey-Research-Results/2012/07/Global-Workforce-Study-2012

van Bilsen, S., Laeven, R., & Nijma, T. (2013). *Escalerende Garantietoezeggingen* (Netspar Design Paper 21 (Dutch)).

van Bilsen, S., Laeven, R., & Nijman, T. (2014). *Consumption and Portfolio Choice Under Loss Aversion and Endogenous Updating of the Reference Level* (Netspar Discussion Paper, 11/2014–048).

van Bilsen, S. & Bovenberg, A. L. (2016). *Personal Pensions with Risk Sharing: Various Approaches*. Netspar Discussion Paper 12/2016-038, see www.netspar.nl

van Ewijk, C., Mehlkopf, R., S. van Bleeken, & Hoets, C. (2015). *Welke keuzemogelijkheden zijn wenselijk vanuit het perspectief van de deelnemer?* Netspar Design Paper(10).

Van Rooij, M., Lusardi, A., & Alessie, R. (2011). Financial Literacy and Stock Market Participation. *Journal of Financial Economics, 101*(2), 449–472.

5

Extending Working Lives

Gerhard Naegele and Jürgen Bauknecht

Introduction

Demographic developments result in a growing number of older workers and pensioners, as well as a remarkable extension in residual life expectancy. On the one hand, this poses challenges to the labour market, and

---

This article is based on MoPAct WP3 reports from Mikkel Barslund | Jürgen Bauknecht | Karolina Beaumont | Miroslav Beblavý, | Inga Blaziene | Carina Bössing | Andreas Cebulla | Nicolas Contrera | Mirko Di Rosa | Katrin Gasior | Vera Gerling | Hanna-Stella Haaristo | Nathan Hudson-Sharp | Elena Jarocińska | Arthur Kaboth | Ingel Kadarik | Veronika Khýrová, Marja Krumina | Reelika Leetma | Karolien Lenaerts | Katja Linnenschmidt, Elisa Martellucci | Märt Masso | Julia Moskvina | Gerd Naegele | Kirsti Nurmela, Liina Osila | Andrea Principi | Ana Ríncon-Aznar | Agnes Romanini | Iryna Shuvaieva, | Marco Socci | Izabela Styczyńska | Anna-Elisabeth Thum | Niko Väänänen, | Lina Vaitkute | Lucie Vidovićová | Marten von Werder | David Wilkinson | Asghar Zaidi | Eszter Zolyomi | Karolina Zubel

G. Naegele (✉)
Research Association for Gerontology, TU Dortmund University, Dortmund, Germany
e-mail: ffg@institut-fuer-gerontologie.de; https://www.tu-dortmund.de/uni/en/Home/

J. Bauknecht
Fliedner Fachhochschule Düsseldorf, Düsseldorf, Germany
e-mail: bauknecht@fliedner-fachhochschule.de

A. Walker (ed.), *The Future of Ageing in Europe*,
https://doi.org/10.1007/978-981-13-1417-9_5

on the other hand, it threatens the long-term financial sustainability of social security systems. In response, European policymakers have implemented reforms in their respective pension, retirement, and labour market policies. These include (among others) higher legal retirement ages, the abolition of early retirement pathways, and the strengthening of active labour market measures (e.g., Hofäcker et al. 2016). Overarching political objectives are to postpone retirement and increase both employment rates and the work volume of those in employment. This was the point of departure for MoPAct work package 3 (WP3) and the following two overarching research questions:

1. How can working lives be extended (and work volumes raised) while raising the employment rate of older workers below and above legal retirement age, including the unemployed, each aided by lifelong (vocational) learning (LLL)?
2. What are socially innovative, effective, sustainable, and transferable private and public sector measures and strategies that enable and encourage older workers (and the unemployed) to both stay in (full-time) employment longer and intensify LLL?

The Key Challenges

Population ageing has led to severe and numerous societal, political, and economic challenges. However, as our research has reconfirmed, strong differences between 'early-' and 'late-mover' countries have to be taken into account when looking for solutions.[1]

Major societal and political risks can firstly be seen in the dissipation of unused human capital. Despite promising success rates in many countries and industries, the employment rates of older workers are still low, particularly in Central and Eastern European (CEE countries), whereas on the other side in some countries/sectors there are increasing skill shortages. Respective risks might arise in the wake of weaker economic growth and declining wealth and linked generational conflicts in the distribution of public goods. Furthermore, continued early and involuntary exit from paid work reduces income below and above legal retirement age and often exacerbates financial problems in old age. As a consequence, poverty and social exclusion in old age particularly among economically vulnerable groups might increase and enhance (old and new) social inequalities. This is particularly true for CEE countries.

Economic risks refer to both the financial sustainability of welfare regimes in general and the so-called great generational contract. Currently, many European Union (EU) Member States are faced with increasing financial pressure on their social security systems, particularly on pension schemes based on a pay-as-you-go principle. This applies even more so against the backdrop of high public debts in most EU countries, binding an increasing share of expenditures to interest payments in the future. However, the old-age dependency ratio tends to overestimate the fiscal costs of population ageing, because it merely depicts the age profile, whereas costs can be influenced by measures promoting later retirement or improving the human capital of older workers. At the macro level, furthermore, ageing societies with widespread early retirement are often characterized by a weaker age-specific demand, resulting from lower disposable incomes in the older age groups.

Research Activities and Objectives

Our research was conducted between March 2013 and February 2017. In all, 13 EU Member States and 8 research bodies were involved deploying a total of 42 researchers, under the scientific direction of TUD as the work package leader. The research was based on a mix of methods, including a literature review, secondary analyses of survey data, reviews of national policy documents, more than 155 case studies of 'promising (good) practices' mainly at the meso (mostly company) level, which served for 21 more in-depth case studies (Bauknecht & Naegele 2015), and a huge number of individual and focus group interviews with older workers, experts with different professional backgrounds, and representatives of social partners, supported by one external expertise. In all, ten major deliverables were produced (comprising more than 2.200 pages of text) http://mopact.group.shef.ac.uk/.

Our research started with a mutual agreement regarding criteria, mainly in terms of 'social innovations' (Heinze and Naegele 2012) and 'promising (good) practices' (Naegele and Walker 2010) used by WP3 research partners as points of departure for their own research activities.[2] Based on this, the main actor groups responsible for realizing the extension of working lives and LLL were identified, followed by a comprehensive policy mapping according to respective responsibilities and spheres of activity. Extensive empirical fieldwork, including document reviews, case study research, interviews, focus groups, and secondary analyses, followed in the main phases of the project. Thus, the basic knowledge was created for both evaluating promising (good) practices and developing an action plan including strategies to extend working lives/volume and LLL. The work was done separately for each country involved. The results were merged and integrated into a final summary report.[3]

The following main actor groups were addressed: older workers, national and public policies/policymakers, public and private employers, social partners, other stakeholders of relevance (like chambers of commerce and national agencies), NGOs, and older persons' representatives.[4] From the spheres of activities identified, those taken up are ones which (both directly and indirectly) affect (positively as well as negatively) chances for extending working lives and LLL: adapting public retirement policies (raising legal retirement ages in combination with financial incentives and closing early labour exit pathways), partial retirement/partial pension schemes, promoting work after retirement ('silver work'), promoting self-employment, anti-age discrimination legislation, protection against dismissals, wage subsidies, active labour market policies for the unemployed, health protection, prevention of disability pension receipt, work-life balance and life-course-oriented policies, socially innovative examples for linking informal and formal work, and information and counselling policies (country targeting national comprehensive programmes and companies targeting strategies).

Respective approaches and activities were evaluated, among others, supported by mutually agreed socio-political assessment criteria with cross-country significance. These were, for example, strengthening generational solidarity, alignment to the interests of the older workers concerned, and avoidance and reduction of social inequalities.

The National Policy Level and Recommendations for an Action Plan

The following state-of-the-art summary refers to those spheres of activities which have been identified as the most promising for extending working life/volume and intensifying LLL, and which at the same time fulfil the prerequisites of being socially innovative good practices. This summary does not aim at completeness, but mirrors what is understood as promising (good) practices and social innovations in different national, political, and cultural understandings in WP3 countries. Furthermore, the dichotomization between early and late movers is taken into account, due to a different understanding of the key dimensions, particularly between Western European and CEE countries.

Adapting National Retirement Policies

Raising Legal Retirement Ages in Combination with Financial Incentives/Disincentives

In all countries studied, stepwise increases of legal retirement ages have been adopted.[5] Results show that higher legal retirement ages, accompanied by pension deductions for those retiring at an earlier retirement age, can lead to longer working lives. Yet, there are remarkable inter-group differences, with weak effects amongst those with low formal education.

Closing/Complicating Early Exit Pathways/Retirement Schemes

Many early exit pathways below legal retirement age have been closed in countries studied, or entry barriers have been raised. This mainly refers to unemployment and disability schemes, but also to special early exit pathways for women. However, there are still some remarkable pathways open, for example, those based on health reasons, particularly for workers in arduous or hazardous jobs, for example, in Belgium, Germany, Finland (from 2017 onwards; '*Years of Service Pension*'), Italy, Poland, and Spain (Friedenstab, Schulz-Weidner and Väänänen 2015).

Generally, employment rates for 55–64 year-olds have risen since the beginning of the crisis (2007–2013) in the EU-27 and EU-28. Figures for 2002–2013 and the age group 50–69 show a similar picture (Peschner 2014). However, solid evaluation criteria in the form of quantitative employment effects measured at best in *real* activity rates are needed. That is, the timing of the labour exit is crucial and not the timing of the start of formal retirement. Furthermore, cohort effects and demographic effects should be taken into account[6] in order to avoid overestimations of policy effects. Such results are only available for a few countries, apart from simulations, but are largely absent for CEE countries. A (rather small) rise in actual retirement ages can be confirmed. Yet, the question is to what degree is the rise in activity rates caused by political reforms and to what degree by macroeconomic circumstances (Eurofound 2011, 2014). If the state of the economy is a crucial factor, then the current development should not be mistaken for the irreversible success of national policies. Furthermore, policy effects are dependent on existing alternative exit pathways out of paid work. Closing one pathway while there are numerous others open may merely have a weak impact, while closing the last attractive pathway out of paid work is more effective.

We found a remarkable linkage to youth unemployment in Southern Europe. Here high (youth) unemployment rates have not stopped these countries' efforts to facilitate higher older worker employment rates and longer working lives, whereas one might assume that in such situations some societal groups erroneously demand that older workers free up jobs for the young in order to reduce possible generational conflicts.

Retirement policies mirror the dichotomization in early- and late-mover countries, which can partially be understood against the background of the prevailing retirement policies in the past. CEE countries emerged from a pre-1990 tradition of low legal retirement ages and special low retirement ages for women and for special occupational groups such as miners and police officers. These low legal retirement ages do not merely reflect a respective tradition. Another rationale is based on the worse state of the health of older CEE citizens and therefore their lower chance of employability.

On the other hand, particularly in CEE countries financial incentives seem to have a strong influence on realizing longer working lives, measured in lifetime pension benefits. Based on OECD pension models, d'Addio et al. (2010) show positive effects of working between 60 and 65 on individual pension wealth (this is true for countries such as the Czech Republic and Poland), and despite comparatively low residual life expectancies there. Work until 65 can be highly expedient in financial terms, ideally combined with sufficient safety-net schemes for those who cannot work for one reason or another.

Deepening Social Inequalities

Not all groups of workers are equally affected by retirement policy adaptations (Hofäcker et al. 2015). High-skilled employees in favourable working conditions and with high salaries have few problems remaining in employment, whereas other groups face multiple barriers. These include low-skilled workers with unstable careers, low incomes, and low pension entitlements. Partly they have to work longer in poorer health and under poor working conditions to ensure a sufficient pension, which is not viable for some (Hess et al. 2016b; Hofäcker 2015). This is confirmed by earlier research documenting socially selective developments of delayed retirement, favouring those with higher pension entitlements or savings, with marketable skills, better health, and higher residual life expectancy (van Solinge and Henkens 2009). Our research confirms that this group has a higher chance of delaying retirement (based on activity rates published by Eurostat for 2015). This is also true for many so-called silver workers (for Germany, see Deller and Maxim 2010; Naegele 2017, for Europe, see also Dubois and Anderson 2012).

Lessons To Be Learnt

Pension scheme adjustments via higher legal retirement ages and the closing of early exit pathways are standard measures to meet the common demographic challenges. However, the results are only partially convincing. Although incentivizing reforms are needed, their impact is weak if initiated in an isolated manner. Corresponding supportive strategies at the meso level of stakeholders and particularly companies are crucial. Workability and employability need to be the fundamental points of departures, including for policy approaches. Financial incentives as isolated measures show the weakest effects, particularly among those objectively bearing the highest financial need to work longer, so that the lack of supportive strategies will increase social inequalities.

There is empirical evidence for rising social inequalities. Since employability differences between status groups can be expected to persist, sensible reforms should follow the guideline 'work for those who can, social security for those who can't'. To increase positive effects of pension reforms in low-socio-economic status (SES) groups, enabling strategies at the policy level as well as the level of employers are necessary, for example, in the fields of health, skills, and the reconciliation of paid work and private life. Pension reforms linking retirement ages to developments of residual life expectancy need to be evaluated with respect to inter-group differences in functional health and residual life expectancy.

Partial Retirement/Partial Pension Schemes

In the past, many EU countries did not permit people to simultaneously remain in paid work above a certain hour/earnings level and draw pensions below official retirement age. In the meantime, most countries are revising their policies partly with the aim of extending working lives (e.g., Belgium, the Czech Republic, Finland, Germany, and Italy). We identified a few innovative examples, such as in Denmark the Voluntary Early Retirement Pension (VERP, also termed 'efterløn') (Barslund et al. 2017, in Italy and Spain, job sharing schemes between older and younger workers).

However, our research emphasizes the question whether partial retirement schemes increase the number of working hours until labour market exit. Research shows vastly negative effects (Bauknecht et al. 2016; Hess et al. 2016a, b; Cahill et al. 2014; Gielen 2009; Graf et al. 2011). The reason is that older part-time workers are those who would have worked full-time if the respective (subsidized) part-time positions had not been available, and are only to a lesser extent workers who would have quit their job completely if the scheme had not have been available.[7] Generous subsidies for part-time work obviously attract too many workers who take the 'good deal' instead of working full-time, although they would have been able to do so.

Lessons To Be Learnt

The disappointing effects mentioned do not speak against the concept of partial retirement as such. Rather, the original idea is worth recollecting: a suitable instrument to raise both quality of work and quality of life among older workers. In this regard, a clear backlog-demand for intelligent combinations of pension receipt with paid work is given. Phased retirement schemes were once primarily developed as explicit gerontological and/or socio-political measures, to improve health, reduce workloads, and to better adjust to 'real' retirement (Bäcker and Naegele 1993). These dimensions are still current; however, they are not primarily for extending working lives, but rather for working longer voluntarily. Following this route, potential for improvement can be found in both incentive structures and employers' attitudes and practices. Collective and/or company agreements could be a promising point of departure (Hess et al. 2016a, b). To meet the needs of the real 'problem cases', micro-targeting of (subsidized or not) old-age part-time work on groups with a high labour market exit hazard under the non-availability of part-time work is promising. Thus, older workers with low physical or mental health or with considerable time constraints, due, for example, to care responsibilities, could be better addressed (Eurofound 2016).

Work Beyond Legal Retirement Age (Silver Work)

Work beyond legal retirement age in principle is permitted in all EU Member States, perhaps more so in some countries financially hampered due to earnings ceilings. Currently, it is more and more common among EU retirees, countries with the largest increases including Finland, Germany, Lithuania, and the UK. Our research did not find evidence for the assumption that retirees would compete with non-retirees for scarce jobs. As a rule the working volume of silver workers is lower than the legal retirement age since it is predominantly performed as part-time employment or minor employment. Official estimations show employment rates between 5% and 10% among pensioners. Working retirees are a heterogeneous group (Dubois and Anderson 2012). From Germany it is known that qualified older workers are overrepresented (Naegele 2016), while from the UK the opposite is reported (Lain 2012). For several countries, it can be expected that declining net replacement rates of public pensions will force more low-skilled retirees into employment, leading to a further heterogenization in terms of silver workers' SES.

However, high shares of undeclared work are assumed. The same is true for work in atypical employment situations, such as in short temporary contracts or as fake self-employment. In the countries studied, declared/standard work applies only for a minority of work beyond the legal retirement age.[8] Silver work can prolong working lives. Yet, silver work is often not just a prolongation of the former employment relationship, but, if motivated by financial necessity, is undertaken in new and low-paid jobs. Thus, silver work can be useful, but close monitoring of causes and conditions is necessary to establish if it is a socio-politically convincing instrument.

Lessons To Be Learnt

Currently, silver work is an option for a small share of those exceeding legal retirement age. Yet, it is of increasing relevance for both an active, satisfying and socially included life in older age, but also for filling incomes gaps caused by declining net replacement rates. The latter seems to grow strongly in EU Member States and should, at best, be avoided by pension adequacy measures, especially since work for the latter reason has a higher probability of being the type characterized as unhealthy. Although the forms of motivation for remaining in paid work beyond legal retirement age are multifaceted, from a normative point of view one can clearly distinguish between two kinds of silver work. First, there is what one might call 'voluntary' silver work by those enjoying their jobs with attached tasks and social contacts and taking the additional income as a positive yet not necessary side effect. This kind of silver work can be considered as clearly positive and expected increases in coming years are a positive development, especially if employers are not forced to retain such workers but may decide to keep or hire them as they wish. Second, there is 'involuntary' silver work due to financial necessity, which can be considered a symptom of mistakes either by the individual or by policies. Nevertheless, given current circumstances it can be expected that both kinds of silver work will be increasingly widespread. If silver work is caused by financial necessity, it is crucial to develop strategies to enable those affected to work in a way which is not further harmful to their health.

Promotion of Self-Employment

Self-employment is seen in many countries as an instrument to be used to aim at both overcoming unemployment and extending working lives. Although there are always risks in self-employment, it is often argued that it is still a better option than unemployment. Similar options are seen in social entrepreneurships or mentors (older workers) supporting and encouraging younger people in self-employment. However, promising measures to foster older-people's self-employment are hard to find, on the macro level as well as the meso level. In most countries, special programmes to support older workers starting their own business exist as a rule for both the employed and unemployed, and in most cases are not limited to certain age groups. An innovative model of self-help and intergenerational cooperation has been run in the Czech Republic: '*We do business TOGETHER*' [*Podnikáme spolu*]). This project brought together unemployed young and older women to introduce them to the possibility of self-employment and equip them with all the necessary knowledge to start their own businesses.

Promotion of self-employment amongst older people can be regarded as an innovative option when it comes to extending working lives, but it bears many motivational barriers, particularly in late-mover countries. These mainly refer to lower social/human capital among many older unemployed, bureaucratic hurdles, no solid financial basis, fear of (new) financial risks in the later stage of life, low knowledge about conducive instruments, and no tailored educational measures. The latter could either be offered publicly or under the scope of collective agreements, as seen in the Netherlands. In early-mover countries, the loss as well as the lack of social security entitlements is one of the most important barriers, according to German (Franke 2012) and Dutch experiences (Gasior and Zolyomi 2015).

To foster self-employment is a recommended option, yet it is still not widespread. There is positive potential for counselling of those interested, as well as in the elimination of all kinds of hurdles. Fair subsidies for new founders are helpful, as well as continued access to acquired social security entitlements. It is helpful to release those unemployed starting their own business from the requirement to prove job search activities to unemployment agencies.

Employment Protection Against Dismissal

Here, national policies are primarily addressed; however, there are many collective and/or company agreements (e.g., in Germany) with similar objectives. In some countries, employment protection does not depend on age but on job tenure; however, both factors are strongly correlated. The instrument as such is discussed critically in some countries due to its ambiguity, which moreover is not completely confirmed empirically.[9] On the one hand, its preventive function is not in question. On the other hand, (too) strict age-specific employment protection might under certain conditions also serve as a risk of reducing hiring and thus worsening the employment chances of older jobseekers.

A remarkable preventive alternative, similar to 'Kurzarbeitergeld' (short-term-allowance, which has helped many German workers stay in their jobs during the economic crisis), is reported in Italy as being worth further examination to see whether it can be transferred to different national environments: the Income Redundancy Fund (*Cassa integrazione guadagni ordinaria,* or *Cigo*) or Extra Income Redundancy Fund (*Cassa integrazione guadagni straordinaria,* or *Cigs*) provides income compensation to workers within a short-term and/or structural firm crisis who have a high risk of being fired (Principi et al. 2012).

Although the positive impact, particularly for older workers, can be confirmed (although with weak empirical evidence), negative effects for both those who are just short of entitlement age and for older jobseekers have to be considered (OECD 2012). Furthermore, there is a de facto employment protection through tailor-made qualifications which has a high value for companies (e.g., experience, expert knowledge) which cannot be replaced simply.

Anti-age Discrimination Legislation

Due to the EU directive, anti-age discrimination legislation is widespread. However, the empirical basis concerning its impact is weak. Although (subjective) age discrimination in practice is empirically evident (e.g., higher in the Netherlands, lower in Denmark), research on the legislation's effects is scarce. Direct effects have not been reported. So far, it is not clear to what degree employers, for example, behave differently due to anti-age discrimination legislation. On the other hand, indirect effects, the power of legislation's norm-setting effect, should not be underestimated. Even though discrimination could be hard to prove in practice, the implementation of anti-age discrimination legislation provides a clear sign to various stakeholders concerning the undesirability of such discrimination.

Wage Subsidies

Wage subsidies aimed at making older workers more attractive to employers are widespread among the countries studied and have proven to be a successful instrument for sustainable integration. They range from direct subsidies for older workers, as in Belgium, to lower social security contributions (in Italy, the Netherlands, Poland, and Spain), to the complete abolition of social security contributions (Italy). This is also true in the case of hiring workers in pre-retirement age (in Poland for the over 50s as part of the *Solidarity across generations* programme) (Bauknecht and Styczynska 2015). A German programme is directly targeted at newly hiring older jobseekers (*Einstiegsgeld*). In the Czech Republic, employers get subsidies for newly hired older workers or new non-bureaucratic *Kurzarbeitergeld* (short-time allowance) to alleviate problems caused by the Russian embargo. In all, it can be shown that subsidized employment decreases the duration of unemployment. However, Italian experiences show the negative effect of subsidies that mainly subsidize temporary and atypical jobs.

Wage subsidies are crucial for raising the chances of reintegrating the older unemployed. Tax reductions can have similar effects. Wage subsidies, and so on, should both be well targeted (particularly at those with the lowest probability of re-employment) and being of short duration. Hiring vouchers can be an alternative to wage subsidies (Brown et al. 2011). Furthermore, close monitoring is crucial due to possible deadweight effects. From a macro as well as a meso perspective, it is necessary to closely monitor the costs and benefits of subsidies since subsidized hiring can reduce unsubsidized hiring or replace unsubsidized workers (Paggiaro et al. 2009; Rotger and Arendt 2011).

Active Labour Market Policies for the Older Unemployed

National policies can also aim at improving the reintegration prospects of the unemployed, by raising their employability. This is particularly true for preventive measures, such as in the fields of health promotion, rehabilitation, and the improvement and updating of skills. Furthermore, our research identified measures targeted at the reintegration of the unemployed with special handicaps (e.g., disabled, long-term sick, and unemployed). The latter is also partly related to health/disability and rehabilitation policies. In some CEE countries, activating programmes have partly been subsidized by EU funding (ESF) and are thus of limited duration, although successful. The most remarkable examples of good practices can be found in Denmark: the activation of (designated) unemployed people starts early, with municipalities obliged to provide senior jobs in the public sector to older unemployed (55–59) who have exhausted their unemployment benefits and are entitled to *VERP* when they are 60. *Senior jobs* are on normal employment terms. However, there is no requirement that individuals must be offered a job corresponding to their qualifications and previous career.

Early activating approaches for (older) workers at risk of becoming unemployed have proven to be successful, also to avoid follow-up risks like loss of functional and extrafunctional qualifications and motivation. Conducive active employment strategies should focus more on enabling strategies instead of exclusively on compensatory policies. The 'Danish model' might be seen as exemplary. Early investment in the employability of workers with acute or latent unemployment risks is extremely helpful. Cost-benefit studies are needed to confirm that the costs of paying unemployment benefits in many cases surmount expenditures better spent earlier for activating programmes. Particular efforts (as well as methods) should be dedicated to older long-term unemployed. To use the local/regional level for activation strategies also seems to be a further good guarantor for success.

Health Protection, Promotion, and Prevention of Disability

Occupational Health Policies

In many national programmes (e.g., the Czech Republic, Poland), the issue of healthy ageing at the workplace plays an increasingly prominent role. In early-mover countries, this is also true for newly introduced collective and/or company agreements. Consequently, both creating awareness and changing health-related attitudes among older workers have become significant tools against the backdrop of a growing number of older workers with health issues across many EU countries. In this context, mental health and/or muscular/skeletal problems are of major significance. This could be supported by a growing positive health consciousness among older workers themselves and their preparedness to invest in their own employability, in areas such as nutrition, gymnastics, healthy lifestyles, sometimes as 'good practice' also supported by companies. In this context, significant drivers are financial incentives for employers to invest in a healthy environment, good health conditions, and in implementing sustainable age management.

Among our partners, Finland has been a pioneer in occupational health policies. Germany and Denmark seem to have become pioneers with their new focus on mental health (stimulated by strongly rising numbers of sickness days and disabilities due to mental health problems). The Netherlands is well known due to the country's remarkable efforts to curb easy and comfortable employment exit via its disability scheme, which is also true for the UK with its successful (in terms of disability numbers) sticks-and-carrots approach to individual barriers to re-employment. Compared to this, in CEE countries as late movers, issues of *healthy ageing* have been taken up comparatively late. In the Czech Republic, they have been fostered by EU membership, while in Poland the 2008 programme *Solidarity across Generations* incorporated the concept of healthy ageing for the first time; this comes against the backdrop of health as a very important barrier to longer working lives in Poland and the WP3 countries with the highest risk of diseases among adults in the countries incorporated by *SHARE*. Furthermore, in Poland ill health is the main reason for exiting the labour market early (Żołnierczak-Zreda 2015). Similar information comes from the Czech Republic which also has high degrees of chronic illnesses and disabilities among older workers. Nevertheless, the relatively bad health status of older workers in CEE countries should not (only) be misinterpreted as being a result of political neglect. Health is affected by long-term factors and CEE countries' current older workers spent a significant share of their lives in the pre-1990 period characterized by worse conditions in both the private and occupational realm.

Disability and Rehabilitation Policies

Disability policies in EU countries have traditionally focused on social insurance security, resulting in considerable increases in the numbers of disability benefit recipients. Efforts to shift to more work-oriented policies over the last decade suggest that fundamental disability reforms, if done well, can lower projected long-term costs for taxpayers, make the job of disability administrators less difficult, and, importantly, improve the short- and long-run job opportunities of people with disabilities. Positive reform experiences in Germany and the Netherlands (and Sweden) point to the considerable potential for keeping workers with disabilities employed, by either eliminating or delaying their movement onto the long-term disability rolls (Zaidi 2017).

Lessons To Be Learnt

Health reasons are still the most significant barriers to employment or re-employment. Therefore, health prevention and promotion are still very important and require an even stronger emphasis in policies. This is particularly true for CEE countries. In general, new health risks such as increasing mental health problems and the ongoing digitization of work have to be taken into account more seriously. A special focus has to be laid on the long-term unemployed; respective endeavours are still too little developed. Health policies targeting older workers suffer from the fact that health policies are long-term oriented, and due to long delays between cause and effect health promotion programmes often come too late. There should be a stronger focus on perceiving younger workers as tomorrow's older workers. Preventive approaches from Finland and the Netherlands where employers have to bear parts of the cost of workers leaving via the disability route could serve as models, but effects on hiring should be monitored.

Our research confirms that it is possible to balance the competing goals of providing social insurance against adverse health shocks for older workers and to maximize their work effort with and without disabilities during late careers. The strategic change in national disability policies in some EU Member States from financing the consequences to the direction of prevention of disability can be regarded as a successful approach not only with high potential for innovation but also as a future-oriented investment into the manpower of an ageing workforce. Disabled older workers could be better integrated by separating disability from unemployment. Furthermore, the preventive potential is higher. Such reforms are not necessarily problematic from a distributional perspective: Disability pensions and progressive unemployment insurance can be used to protect those with low earning capacities close to pension age. Positive reform experiences point to the considerable potential of keeping workers with disabilities employed, and either eliminating or delaying their movement onto the long-term disability rolls. Efforts to shift to more pro-work policies over the last decade suggest that fundamental disability reforms can lower projected long-term costs for taxpayers.

Work-Life Balance and Life-Course-Oriented Policies

Work-life-balance policies react both to the increasing importance of private life dimensions in work and retirement (as explicitly taken up in the concept of workability and employability) and to changes in work-related private needs and motivations which to a great extent result from overall social, demographic and political changes (Naegele et al. 2003). Quality of work and quality of life (as conceptualized as 'social environment' and 'family' in the 'house of workability') (Ilmarinen 2005) are increasingly important dimensions in the process of both individual retirement and decisions to work longer.

Life-Course Orientation in Employment Policies—To Make Working Life More Flexible

Although life-course orientation has been regarded as highly innovative for years, especially through its focus on prolonging working lives (Naegele et al. 2003), so far it is only represented in a few approaches at the company level (see Chap. 6). The concept itself has its origins in the assumption that in order to change corporate life one has to understand human life, since the latter gives individuals 'the strength to cope with everyday chores and provides a basis for a good retirement – or not' (Ilmarinen 2005: 101). The Dutch approaches of *Levensloopsregeling* (de Grip et al. 2010: 6; abolished in 2012; Delsen and Smits 2014) and

Vitaliteitsregeling belong to the best-known models, next to the German approach of *working time accounts* regulated in collective and/or company agreements. However, in many cases all have been/are misused as pathways into early retirement. Furthermore, innovative models can be found in Belgium as the *career break* and the *end-of-career time-credit* system. One of the most important fields for life-course policies in the future will be elderly care.

The Case of Reconciling Work and Care

Explicit life-course approaches are different from those aiming at *risky* live events. Regarding the latter, one case has gained outstanding significance in our research: Reconciling work and care, which has gained increasing importance for (early) retirement decisions (Reichert et al. 2018). Although of increasing significance, this risk is mostly regarded and treated as an isolated social event and not part of an overarching life-course context. Rising female older worker employment rates due to cohort effects will, in particular, necessitate innovative solutions. Possibly care leave can be part of the solution, but such policies carry the risk of also reducing labour market attractiveness for some groups, primarily (older) women. This general neglect is confirmed by a recently published EC report stating that 'too little attention has been paid to the role of the carer' since this has been 'rarely recognised as a problematic issue as such. In the majority of countries, LTC policies are geared to the dependent person and overlook the carer's situation' (Bouquet et al. 2016: 8).

Informal Work/Social Volunteering in Combination with Paid Work

No case of promising (good) practice could be identified in the countries we studied where paid work and social volunteering and/or civic engagement in later stages of working life is incentivized either by public policies or by collective and/or company agreements. This mirrors well-known differences between early- and late-mover countries (Ehlers et al. 2011).

While social volunteering in older age is usual in the Netherlands, it is particularly unusual in CEE countries, mainly due to the absence of historical experiences. One barrier to volunteering could be full-time work. Theoretically, opportunities for reducing working time in later phases of working life for social volunteering are provided through part-time/partial pension schemes.

Lessons To Be Learnt

So far a life-course orientation in public policies is not well recognized (except in the Netherlands and Germany, but here with ambiguous impact and often misused for early retirement). The idea embodied in the former Dutch *levensverloopsregeling* still has innovative potential worthy of imitation. There is also a strong potential for life-course policies in collective and/or company agreements but these are used too rarely.

Informal care for the elderly is a severe employment obstacle for mainly elderly women and thus as a significant labour market challenge. Further recognition is necessary through public policies for effective and sustainable measures of reconciling work and elder care beyond financial compensation, including removing administrative and financial barriers. More true care leaves are needed, as well as factual support by key actors at the meso level. Reconciling endeavours so far are more effective in countries in which various schemes of part-time and flexible working time options are widespread. However, policies are also addressed: Fragmentations of benefits and services between branches of social security systems, especially between health and social assistance (as in many EU Member States), have to be overcome.

Awareness, Information, and Counselling Policies

Country-Targeting National/Public Programmes

We found innovative examples in different countries (the Czech Republic, Estonia, Finland, Poland, the UK). Also worth mention is the Spanish programme being developed on the *Global Strategy for the Employment of Older Workers* (2012–2014). The most significant elements are (1) promoting healthy and secure working conditions (based among other factors on vulnerability analyses and training needs studies); (2) enhancing flexibility within companies, especially employers of older workers; (3) reshaping public employment services according to the demands of older unemployed workers to improve their employability and to establish better links to employment options; (4) improving self-employment, particularly as a 'joint venture' between young and older unemployed; and finally (5) combating age discrimination within companies.

Company-Targeting National Strategies

The potential of good practices as a role model providing crucial information has been proven repeatedly and particularly in late-mover countries. It definitely should serve as a starting point for further information measures and approaches and might serve as models for late-mover countries. However, they need to be continually updated and evaluated, and positive framework conditions should be taken into consideration and disseminated. In many countries information and awareness programmes used to inform and convince key actors to invest in the manpower of older workers and/or to recruit older jobseekers and/or to retain older workers longer in working life could be found. Besides legislation, in nearly every country we could find positive examples of employers who have been successfully given tools and information to deal with an older workforce (D'Addio et al. 2010). Outstanding examples include two German initiatives *The demography network* (ddn) and *New Quality of Work* (INQA) which can be regarded as highly innovative.

Lessons To Be Learnt

National information and awareness programmes have proven very helpful in strategic terms, for example, in creating and disseminating 'demographic consciousness' among key actor groups, and supporting and initiating measures on the meso/company level. Further supportive action could include strong anti-discrimination/equality legislation in order to induce a cultural as well as an economic shift that accepts, integrates, and strengthens an ageing workforce (Lain 2012). However, a thorough scientific evaluation so far is absent and urgently needed. National strategies targeting the company level are most helpful particularly in late-mover countries. The learning potential of promising (good) practices as role models providing crucial information has been repeatedly proven. It should thus be used more often and in a more strategic way. Furthermore, there is an urgent need for continual updating and evaluation. The specific country examples show that initiatives can successfully be integrated in respective national employment policies.

The Meso/Company Level

Different Understanding of Social Innovation

Located at this level are employers as well as social partners. We explicitly focused on the company level as the most significant level to prolong working lives, taking the criteria of social innovations and promising (good) practices as prime points of departure. Not surprisingly, the term 'social innovation' was understood differently in different national contexts. There is no common comprehension of the term, particularly at the company level (Chap. 1). In early-mover countries, innovative measures are closer to the classic definition of innovation, including unique social inventions never put into practice before, whereas in late-mover countries social innovations are closer to the wider definition including measures put into practice before elsewhere—albeit always at least slightly differently—but not put into practice in the specific context.

Spheres of Activities

Our research revealed a differentiated picture of the age management-landscape assessed as socially innovative. This partly mirrors the respective state of thematic awareness among employers, their associations, and other key actor groups in the countries involved. The dimensions of work-life balance, quality of work, and life as well as life-course orientation are gaining importance with a focus on early-mover countries. In all, the 155 identified cases of promising (good) practices depicted can be systematized as follows[10] (see also Naegele and Walker 2010):

- retaining and promoting health
- to promote working longer/work after retirement
- adjustment working time and flexible working time schedules
- work load reduction and ergonomics
- special measures for overloaded workers (e.g., shift workers)
- career development and planning
- work-life-balance policies
- life-course-oriented personnel policies
- aged-mixed teams
- reconciliation of paid work and care
- overcoming unemployment/active integration measures
- recruitment
- gliding retirement schemes
- preparation for retirement
- fostering self-employment
- awareness campaigns
- knowledge transfer, skill promotion, and LLL.

Drivers of and Barriers Against Extending Working Lives of Older Staff Members

In general, successful socially innovative approaches are strongly dependent on the sector and the organization size. The latter is caused by differences between larger and smaller companies, resulting in the special need of SMEs since (1) their capabilities are more restricted compared to large enterprises and (2) however, their total number of workers is high.

The main **drivers** were (1) external circumstances (e.g., financial situation, lack of skilled labour, and macroeconomic situation); (2) in many cases and particularly in CEE countries *EU funding*; however, after the end of funding the probability that measures are stopped is high. (3) Legislation on the macro level as an initial impetus (e.g., Italy: 'Fornero reform' increasing the legal retirement age). (4) Collective and/or company agreements have served as a catalyst, and if so have proven to be a good basis for sustainability.

Our case study research has shown that close and informal contacts between crucial actors (HR departments, CEOs, workers' representatives, and the older workers) are good prerequisites. Doubtless, this is easier to achieve in SMEs. Very often HR departments are identified as the key drivers for organizational efforts in terms of good practice in corporate age management. In SMEs, leadership is possibly the crucial factor. This can be a constellation of actors, yet in numerous cases innovative measures can be traced back to a single person, for example, the CEO or the head of the HR department. Also, cooperation between SMEs can be helpful but often needs external impetus. In all, key actors' attitudes are one crucial determinant for the existence or absence of good practice measures. To inform and to convince these key actors could be expedient. Furthermore, there is a strong influence of company-level path dependence.

Concerning the **sustainability** of measures, company/organization-level measures often result in a 'win-win' situation (at least subjectively). Several measures would not have been implemented to begin with if not all relevant actors had expected net gains. If all relevant actors are content with measures, the probability of reversal is low, at least as long as there is no staff turnover amongst key personnel.

The main barriers were employers' misperception concerning employees' absence records, health, trainability, and productivity, and related to this, no motivation to retain older workers longer. An open question is the influence of rising (seniority) wages with age: In some OECD countries, older employees have higher or markedly higher wages than prime age employees, which is problematic if this does not come with higher productivity (OECD 2011). On the macro level, too high seniority wages as well as too strong employment protection seem to be negatively related to older male workers' employment rates (OECD 2011).

When looking for determinants of retaining and rehiring of older workers by companies in six European countries, secondary analyses,[11] originally stemming from the project 'Activating Senior Potential in an Ageing Europe' (ASPA), suggest that the main drivers for companies' approaches are the implementation of HR measures for older workers to improve their performance, their working conditions, as well as to reduce costs linked to their management. Other drivers are a high share of older workers in the company, medium or large company size, and having experienced recruitment problems. There are clear indications that the possibility of reducing costs due to EWL practices is a key element for companies. In times of crisis, however, even these reduced costs could be 'sacrificed' by companies, as suggested by the generally negative association between approaches and the need to reduce staff levels. There are indications that in the manufacturing sector the qualities, skills, and knowledge of older workers could be more valued and desired than in other sectors. Indeed, the adoption of approaches in this sector is linked to the provision of an accommodating measure as part-time options for older workers, and to high knowledge intensity in the company. There are indications that in the public sector there may be partially different models of EWL practices' mechanisms: there might be less motivation to retain older workers' qualities, and economic problems could have a milder effect on these practices. Some indications suggest that age stereotypes may be present, especially in the service sector.

Conclusions and Overall Lessons To Be Learnt

National policies can only give incentives for longer working lives. The employer level is crucial for enabling older workers to work longer. This research confirms that organizations play a key role in supporting older workers to extend working lives, but need support from national policies as well as from social partners and stakeholders. Particular focus should be on SMEs. In late-mover countries, organizations need greater external incentives and support (e.g., legal and financial incentives, professional advice, and dissemination of good practice) than in early-mover countries where both awareness and knowledge of the state of the art are more common. In both, disseminating good practice is necessary. Particularly with respect to the organizational level, more innovative approaches by different actors (e.g., companies, organizations, and social partners from different sectors with different financial/HR capabilities) are needed. In all, the concepts of *workability* and *employability* are the best points of departure for designing policies and measures.

As the main drivers, HR departments and workers' representatives could be the main targets of external (policy) intervention. Collective and/or company agreements can facilitate measures, and be good foundations for sustainability. However, dissemination gaps raise the issue of social inequalities. The participation of work councils—if available—is essential. Since HR departments have (partly) to justify their activities in front of company/organization leaders, monetary cost-benefit calculations for specific companies are urgently needed but, however, are largely non-existent.

Recommendations aimed at the company level need to be aware of late- and early-mover differentiations. In late-mover countries, the dimensions of health, skills, and working conditions are still the most relevant dimensions of the workability/employability of older workers. Companies, organizations, stakeholders, and so on need more knowledge about preventive strategies on different levels. Also in early-mover countries, health reasons are still of some relevance and workers with health problems are among the typical problem groups.

Life-course-oriented HR strategies can be characterized as innovative (e.g., flexible working time schedules, a new organization of working time over the life course, career planning, life-course-oriented measures to develop and retain skills and competences). Labour mobility amongst older workers between jobs and employers is an instrument conducive to promoting and retaining skills. Furthermore, work-related attitudes, values, and work-life-balance-related dimensions as taken up in the concepts of workability/employability are playing an increasing role, particularly among higher qualified workforce in early-mover countries, and should be used more often as starting points for action. Due to the heterogenization of biographies and the multitude of individual life circumstances, the individualization of pro-workability/employability measures becomes even more necessary, at best embedded in a profound knowledge about company-typical critical work-life stages.

Lifelong (Vocational) Learning

State of the Art

The concept of LLL had already been mentioned and discussed in Europe in the 1970s and afterwards was made more prominent by the 'European year of LifeLong Learning' in 1996. LLL includes formal, non-formal, and informal education. Training which is set up for professional education in institutions (universities, etc.) is categorized as formal education. On the contrary, non-formal education refers to training in institutions that are not mainly responsible for education, for example, companies. Informal learning implies learning through experiences gathered in certain stages of the individual (working) life course. Various results show that learning for the elderly should be adapted to their special needs, for example, it should be application-oriented and need to correspond to the respective utility value. This is particularly true for vocational training aimed at older workers. How to best organize vocational training in methodological terms is seen differently in our countries: On the one hand, findings from Italy and Finland argue that frontal forms of vocational training are ill-suited to older workers. On the other hand, findings from Germany show that older workers dislike non-formal activities, which is confirmed by results from Denmark especially amongst lower educated people.

There are several reasons documenting the increasing importance of both vocational and general LLL in ageing societies. First, against the backdrop of longer working lives and the rapid obsolescence of knowledge, LLL is a significant precondition to maintain and strengthen employability and to keep older workers/an ageing workforce productive. Skills mismatch, in particular, plays an important role. Second, intergenerational exchange is important for companies' competitiveness and sustainable management of knowledge. Third, LLL has a clear life-course dimension: The widespread image that learning is particularly necessary in younger age (which is mirrored in front-biased educational policies in many EU Member States) is losing its meaning. Fourth, LLL keeps older people active, enables them to participate in society, and helps them to stay self-responsible and independent. Fifth, there is a positive correlation with health and well-being, which is especially true for non-formal activities.

Results from Finland demonstrate the importance of LLL. Nearly half of workers aged 50–64 report that it helps to increase their likelihood of maintaining work. Also, positive impacts on extending working lives are reported, whereas findings from Denmark show a neutral effect on retirement decisions. One possible barrier which occurs in the UK is the low investment of employers in further training for older workers as well as the low quality of the training. Moreover, in the Baltic countries, aside from the macro and meso levels, the micro level—motivation of individuals—is an important topic and may be a possible barrier to training. A possible driver that may lead to higher participation rates is paid educational leave for workers in the private sector, as financed by the Belgian government. Our research revealed the following prerequisites and barriers for improving LLL activities and quality:

- Training for low-skilled workers to improve employability is seen as an urgent task both in early-mover and in late-mover states.
- In many countries, age limits have an impact on the decision to participate in further training. In Belgium, evidence for rather widespread participation in informal learning was detected among those older than 50.
- A further approach in CEE countries refers to stronger endeavours by employers. To reduce costs on the employee's side is seen as a necessary incentive to overcome barriers in the Baltic states. In Estonia, for example, it is recommended that employers should provide employees with training based on the interests of the employer's enterprise, carry the training expenses as well as pay wages during the training period. In this context, direct subsidies are recommended as an additional incentive as well as tax deductions or tax exemptions.

- Many country experiences confirm obligations (legal or collective/company agreements) to be a suitable instrument for extending LLL. Collective and/or company agreements should play a more decisive role. However, this necessitates the existence of collective agreements.
- The Italian country report neatly summarizes recommendations to foster LLL, which are valid not only for Italy: (1) financial incentives not only for 'classic' further training (seminars, workshops, etc.), (2) awareness campaigns to encourage older workers' motivation to participate, (3) (to incentivize) companies' support for older workers' training, (4) more information about paid/unpaid educational leave, (5) regulatory systems concerning tax benefits for employers (Principi et al. 2016).

Lessons To Be Learnt

LLL is crucial for the improvement of the workability and employability of an ageing workforce; LLL is also necessary for transferring skills as well as preparing for critical working-life transitions. LLL is of rising importance in all countries, especially in late-mover countries. Here in particular a better subsidized infrastructure for learning and upskilling is needed. In early-mover countries, the clear dominance of the Matthew principle of giving to those who already have is a barrier to targeted LLL.

For all countries, the widespread bias of learning policies towards early stages of life is a second large barrier: 'The vicious cycle between retirement and low training investments in older workers has to be interrupted' (Principi et al. 2016). In ageing societies, this front-bias needs to be overcome.

Collective and company agreements have proven conducive to increasing LLL activities on the meso level and need to be extended (particularly in late-mover countries). Public training leave has also proven to be conducive, as well as working-time accounts if used for training purposes. LLL should be supported through ICT; e-learning should be expanded further. The enormous gap in terms of available public funds between formal and other less formal kinds of learning provision needs to be overcome.

LLL should be adapted to the special needs and interests of older persons, inside and outside of companies. This often implies a more practical focus. The transferability of skills should be fostered, which also necessitates improved recognition of informally acquired skills and experiences. Mentoring-models haven proven to be successful. LLL activities need to be focused on those underrepresented in further training; a special emphasis should be laid on long-term unemployed or older workers with a higher unemployment risk; this is also true for older workers on short-term contracts.

Final Conclusions

We disagree with extremely bleak outlooks based on demographic forecasts. Our MoPAct research repeatedly confirms that policy approaches on all levels, national, meso, and individual, have the potential for both sustainable and socially innovative strategies to promote active and healthy ageing in employment. Demographic challenges should be met at best through adequate adaptations of the labour world to make the ageing of the population as well as the workforce an economic resource. This can be realized through longer working lives, higher older worker's employment rates, and higher numbers of weekly working hours, supported by intensified LLL. Thereby, social security systems as well as the underlying generational contract could be adjusted and stabilized, while at the same time fiscal and social sustainability might be improved and higher tax revenues and pensions could be provided. A key prerequisite is to keep an ageing and shrinking population productive and active longer in working and social life, through investment in their human capital. Longer working lives might benefit older workers economically, socially, and intrinsically, while for many it is also beneficial in terms of healthy ageing.

In all, policies supporting ageing individuals to maintain and promote their own potential and human capital are needed—according to the subsidiarity principle (or capability approach). Therefore, conductive and supportive structures in public responsibility are required. There is an urgent need for a stronger focus on the meso level through public policy measures, with a particular need to emphasize SMEs. So far this seems to be underdeveloped and, and if it exists at all, it is often more a side effect than an intentional result. However, national policies can only give incentives for longer working lives. The employer level is crucial for enabling older workers to work longer. Our research confirms that organizations play an outstanding role in supporting older workers to extend working lives, but need support from national policies as well as from social partners and stakeholders. In all, more policy-mix approaches are needed. They have proven to produce the strongest effects. However, their distribution rate is poor. Furthermore, cooperation between research and policymakers has proven fruitful and offers a good basis for sustainable measures. National polices should, if they have not already, promote more older worker research.

Notes

1. Early-mover countries: Denmark, Finland, Belgium, the Netherlands, Germany, Italy, the UK. Late movers: Spain, and the Central and Eastern European (CEE) countries Poland, the Czech Republic, the Baltic countries Estonia, Latvia, and Lithuania.
2. For this purpose TUD developed a conceptual framework which served as overarching research guidelines for the WP§ countries involved: Naegele, G. & Bauknecht, J. (eds.) (2013): MoPAct: Extending working lives. Conceptual Framework, work package 3, task 1. With support of Mikkel Barslund | Andreas Cebulla | Nicolas Contreras | Katrin Gasior | Reelika Leetmaa | Elisa Martelluci | Kirsti Nurmela | Liina Osila | Andrea Principi | Ana Rincón-Aznar | Marco Socci | Izabela Styczynska | Anna-Elisabeth Thum | Marten v. Werder david Wilkinson | Lucie Vidovicová | Eszter Zolyomi.
3. Naegele, G. & Bauknecht, J. (2017): Strategies ("Action plan") for extending working lives, raising older workers' employment rates and intensifying LLL in later working life. MoPacT final report. WP 3. Dortmund: FFG.
4. In this chapter only the national policy level and the meso level are considered. For more information about the role of the older workers themselves, social partners, other stakeholders and NGOs, see Naegele, G. and Bauknecht, J. (2017).

5. For an overview, see European Commission (2012).
6. As in Germany, with cohort effects assumed to amount to a quarter of the rise (Peschner 2014).
7. A recently completed secondary analysis carried out by Hess, Bauknecht and Pink (2016) in the context of the EWL project concludes: '*(…) the results show that shifting from full-time work to part-time work might lead to earlier retirement and lower labour supply*' (p. 17).
8. See Dubois and Anderson (2012: 51f.) for a complete overview of details of public pensions and work after retirement for 2012.
9. This is also mentioned for (higher) seniority wages in a few country reports (e.g., for the Netherlands) as possibly serving as a barrier for the recruitment of older jobseekers.
10. Bauknecht, J. & Naegele, G. (eds.) (2015): MoPAct. Extending working lives. Best Practice Cases Work Package 3 task 2. Barslund, M., Bauknecht, J., Cebulla, A., Gasior, K., Kadarik, I., Khyrova, V., Krumina, M., Linnenschmidt, K., Martellucci, E., Moskvina, J., Naegele, G., Osila, L., Principi, A., Rincon-Aznar, A., Socci, M., Styczyńska, I., Vidovićová, L., Wilkinson, D. & Zólyomi, E.
11. Principi, A., Bauknecht, J., Socci, M. & di Rosa, M. (2016): Labour demand and longer working lives in Europe: drivers and barriers in companies. In: Report on Secondary Analyses. MoPAct WP3, 29–56.

References

Bäcker, G., & Naegele, G. (1993). *Alternde Gesellschaft und Erwerbstätigkeit, Modelle zum Übergang vom Erwerbsleben in den Ruhestand.* Köln: Bund-Verlag.

Barslund, M., Bauknecht, J., & Gerling, V. (2017). *MoPAct Final Country Report.* Denmark: MOPACT: WP3.

Bauknecht, J., & Naegele, G. (2015). *MoPAct Best Practice Cases.* WP 3.

Bauknecht, J., & Styczyńska, I. (2015). National Policy Report – Poland. In G. Naegele & J. Bauknecht (Eds.), *MoPAct National Policy Reports* (pp. 97–105). Unpublished project report.

Bauknecht, J., Gerling, V., & Hess, M. (2016). *Extending Working Lives Through Flexible Pension Schemes. Case Studies with Focus on Partial Retirement Schemes.* Unpublished project report.

Bouquet, D., Spasova, S., & Vanhercke, B. (2016). *Work-Life-Balance Measures for Persons in Working Age with Dependent Relatives in Europe. A Study of National Policies*. European Social Protection Network (ESPN), Brussels.

Brown, A. G., Merkl, C., & Snower, D. L. (2011). Comparing the Effectiveness of Employment Subsidies. *Labour Economics, 18*, 168–179.

Cahill, K. E., Giandrea, M. D., & Quinn, J. F. (2014). *The Impact of Hours Flexibility on Career Employment, Bridge Jobs, and the Timing of Retirement* (Bureau of Labour Statistics Working Paper, 472).

D'Addio, A. C., Keese, M., & Whitehouse, E. (2010). Population Ageing and Labour Markets. *Oxford Review of Economic Policy, 26*(4), 613–635.

de Grip, A., Lindeboom, M., & Montizaan, R. (2010). Shattered Dreams: The Effects of Changing the Pension System Late in the Game. *fo/CESifo and University of Munich Conference on Empirical Health Economics.*

Deller, J., & Maxim, L. (2010). Activities in Retirement: Individual Experience of Silver Work. *Comparative Population Studies, 35*, 810–832.

Delsen, L., & Smits, J. (2014). *The Rise and Fall of the Dutch Savings Scheme* (Netspar Discussions Papers). https://doi.org/10.2139/ssrn.2414546.

Dubois, H., & Anderson, R. (2012). *Income from Work After Retirement in the EU*. Luxembourg: European Union.

Ehlers, A., Naegele, G., & Reichert, M. (2011). *Volunteering by Older People in the EU*. Dublin: European Foundation for the Improvement of Living and Working Conditions.

Eurofound. (2011). *Living Longer, Working Better – Older Workers in the Recession*. Dublin: Eurofound.

Eurofound. (2014). *Foundation Findings. Work Preferences After 50. 3rd EQLS Policy Brief.* Dublin: Eurofound.

Eurofound. (2016). *Extending Working Lives Through Flexible Retirement Schemes: Partial Retirement*. Luxembourg: Publications Office of the European Union.

European Commission. (2012). *European Employment Observatory Review, Employment Policies to Promote Active Ageing 2012*. Birmingham: European Commission.

Franke, A. (2012). *Gründungsaktivitäten in der zweiten Lebenshälfte: Eine empirische Untersuchung zu Gründern 50plus im Kontext der Altersproduktivitätsdiskussion.* Wiesbaden: Springer VS.

Friedenstab, C., Schulz-Weidner, W., & Väänänen, N. (2015). Tätigkeitsbezogene Frührentensysteme in Europa – die Bewertung besonderer beruflicher Belastungen. *Deutsche Rentenversicherung, 3*, 183–199.

Gasior, K., & Zolyomi, E. (2015). *Identifying and Assessing Structural Drivers of and Barriers to Innovative, Sustainable Strategies for Extending Working Lives and Lifelong Learning on the Demand and Supply Side*. Vienna: MoPAct National Report.

Gielen, A. C. (2009). Working Hours Flexibility and Older Workers' Labour Supply. *Oxford Economic Papers, 61*, 240–274.

Graf, N., Hofer, H., & Winter-Ebmer, R. (2011). Labour Supply Effects of a Subsidised Old-Age Part-Time Scheme in Austria. *Journal for Labour Market Research, 44*(3), 217–229.

Heinze, R. G., & Naegele, G. (2012). Social Innovations in Ageing Societies. In F. Hans-Werner, J. Hochgerner, & J. Howaldt (Eds.), *Challenge Social Innovation* (pp. 153–167). Berlin/Heidelberg: Springer.

Hess, M., Bauknecht, J., & Pink, S. (2016a). *Hours Flexibility and Timing of Retirement – Findings from Europe* (Dortmund, FFG EWL-Working Paper).

Hess, M., König, S., & Hofäcker, D. (2016b). Retirement Transitions Under Changing Institutional Conditions: Towards Increasing Inequalities? Comparing Evidence from 13 Countries. In D. Hofäcker et al. (Eds.), *Delaying Retirement: Progress and Challenges of Active Ageing in Europe, the United States and Japan* (pp. 363–378). London: Palgrave Macmillan.

Hofäcker, D. (2015). In Line or at Odds with Active Ageing Policies? Exploring Patterns of Retirement Preferences in Europe. *Ageing and Society, 35*(07), 1529–1556.

Hofäcker, D., Naumann, E., & Hess, M. (2015). Changing Retirement Transitions in Times of Paradigmatic Political Change: Towards Growing Inequalities? In C. Torp (Ed.), *Challenges of Aging: Retirement, Pensions, and Intergenerational Justice* (pp. 205–226). Houndmills: Palgrave Macmillan.

Hofäcker, D., Hess, M., & König, S. (2016). *Delaying Retirement: Progress and Challenges of Active Ageing in Europe, the United States and Japan*. London: Palgrave Macmillan.

Ilmarinen, J. (2005). *Towards a Longer Worklife! Ageing and the Quality of Worklife in the European Union*. Helsinki: Finnish Institute of Occupational Health, Ministry of Social Affairs and Health.

Lain, D. (2012). Working Past 65 in the UK and the USA: Segregation into 'Lopaq' Occupations? *Work, Employment and Society, 26*(1), 78–94.

Naegele, G. (2016). Der demografische Wandel hat längst die Arbeitswelt erreicht – und was ist passiert ? Ältere Arbeitnehmer als „Produzenten" wirtschaftlicher Potenziale. In Hoose, Fabian, Beckmann, Fabian, & Schönauer, Anna-Lena (Hrsg.), *Fortsetzung folgt. Kontinuität und Wandel in Wirtschaft und Gesellschaft* (pp. 351–376). Berlin: Springer/VS.

Naegele, G. (2017). Altes und Neues zur „Arbeit trotz Rente" – Weiter wachsende Rentnererwerbsarbeit. In Generali Deutschland AG (Hrsg.), *Generali Altersstudie 2017. Wie ältere Menschen in Deutschland denken und leben* (pp. 73–79). Berlin: Springer.

Naegele, G., & Bauknecht, J. (2017). *Strategies ("Action Plan") for Extending Working Lives, Raising Older Workers' Employment Rates and Intensifying LLL in Later Working Life*. Dortmund: FFG: MoPAct final report. WP 3.

Naegele, G., & Walker, A. (2010). In M. Malloch, L. Cairns, K. Evans, & B. N. O'Conner (Eds.)., *The SAGE Handbook of Workplace Learning Age Management in Organisations in the European Union* (pp. 251–268). London: SAGE.

Naegele, G., Barkholdt, C., de Vroom, B., Andersen, G. J., & Krämer, K. (2003). *A New Organization of Time Over Working Life*. Dublin: European Foundation for the Improvement of Living and Working Conditions.

OECD. (2011). Helping Older Workers Find and Retain Jobs. In: *Pensions at a Glance 2011: Retirement-Income Systems in OECD and G20 Countries.*

OECD. (2012). *Thematic Follow-Up Review of Policies to Improve Labour Market Prospects for Older Workers*. Germany. Paris: OECD.

Paggiaro, A., Rettore, E., & Trivellato, U. (2009). The Effect of a Longer Eligibility to a Labour Market Programme for Dismissed Workers. *Labour, 23*(1), 37–66.

Peschner, J. (2014). Structural Drivers of Activity and Employment Rates of Older Workers. *Workshop on Delivering Longer Working Lives and Higher Retirement Ages, OECD & European Commission,* Brussels.

Principi, A., Bauknecht, J., Socci, M., & di Rosa, M. (2016). *Labour Demand and Longer Working Lives in Europe: Drivers and Barriers in Companies.* MoPAct WP3 Deliverable D3.4.2.

Principi, A., Checcucci, P. and Di Rosa, M. (2012). Income from work after retirement in Italy, Ancona, INRCA. Available at http://www.inrca.it/inrca/files/focuson/Income from work after retirement in Italy.pdf.

Reichert, M., Peres-Bialas, J., & Hamblin, K. (2018). *Carers@work: Konflikt oder Chance? Vier europäische Länder im Vergleich*. Wiesbaden: Springer: VS: in Press.

Rotger, G. P., & Arendt, J. N. (2011). The Effect of a Wage Subsidy on Employment in the Subsidised Firm. Presentation at: *The Sixteenth Annual Meetings of the Society of Labour Economists (SOLE)* (pp. 29–30). Vancouver.

van Solinge, H., & Henkens, K. (2009). Living Longer, Working Longer? The Impact of Subjective Life Expectancy on Retirement Intentions and Behavior. *The European Journal of Public Health, 19*, 1–5.

Zaidi, A. (2017). *Health Impairments and Functional Limitations. Comparative Analysis of Employment of Older Workers with Disabilities fur EU Countries* (MoPAct Working Paper, Third Draft).

Żołniercz-Zreda, D. (2015), National Report Poland, Full Report. In Hasselhorn, H.-M., & Apt, W. (Eds.), *Understanding Employment Participation of Older Workers: Creating a Knowledge Base for Future Labour Market Challenges*. Berlin Zolnierz-Zredea: Research Report. Federal Ministry of Labour and Social Affairs (BMAS) and Federal Institute for Occupational Safety and Health (BAuA). BMAS/BAuA.

6

Healthy Life Years and Social Engagement

Dorly J. H. Deeg, Marie-Louise Boertje, and Henrike Galenkamp

Introduction

With life expectancy (LE) increasing, it is important to foster socially productive and satisfying lives for older people, and thus to improve ways in which older people can contribute to society. Social involvement not only contributes to a sense of belonging but it may also foster good health.

All other members of the research group who contributed to the research reported in this chapter: Stella Golinowska, Stefania Ilinca, Guus Luijben, Jan-Willem van der Maat, Jane Matt, Amilcar Moreira, Roberta Papa, Andrea Principi, Roberto Rodrigues, Andrea Schmidt, Katherine Schulmann, Bert Smid, Marco Socci, Agnieszka Sowa, Vootele Veldre, Rauno Vinni, Bram Wouterse.

D. J. H. Deeg (✉) • M.-L. Boertje
Department of Epidemiology and Biostatistics, VU University Medical Center, Amsterdam, The Netherlands

Amsterdam Public Health Research Institute, Amsterdam, The Netherlands
e-mail: djh.deeg@vumc.nl

H. Galenkamp
Amsterdam Public Health Research Institute, Amsterdam, The Netherlands

Department of Public Health, Academic Medical Center, University of Amsterdam, Amsterdam, The Netherlands
e-mail: h.galenkamp@amc.nl

A. Walker (ed.), *The Future of Ageing in Europe*,
https://doi.org/10.1007/978-981-13-1417-9_6

The continued increase in LE is currently driven by mortality declines at older ages. These declines can be attributed partly to better conditions throughout the life course, partly to improvements in medical care. As improvements in medical care are mainly benefitting people with chronic diseases, it is expected that the prevalence of chronic conditions will increase, especially in the older population. By consequence, healthy life expectancy (HLE) may not increase at the same pace as total LE, and ways should be sought in which the health of older people with chronic diseases can be improved.

Therefore, this part of the MoPAct project focuses on social involvement, its driving factors, and its meaning for HLE, especially in older people with health limitations. We examined individual determinants of social participation in older people in poor health, identified strategies—of both NGOs and the government—for fostering social participation, and defined potential gains in HLE according to various scenarios of future increase in social participation.

Changing Demographics

Mortality rates of people aged 50 and over have decreased in all European Union (EU) countries over the past 30 years for both men and women. This can largely be attributed to declining mortality rates from major chronic diseases such as cardiovascular and cerebrovascular diseases, diabetes and cancer (Deeg et al. 2013). This is the result of advances in medical care, healthier lifestyles, and improved living conditions, both at younger and older age. These reductions in mortality rates are increasingly concentrated at older ages and have led to large increases in LE.

While both improvement in living conditions and medical care enable longer lives, the first factor affects longevity in both healthy and diseased older people, and the second factor allows longer lives in particular in older people with diseases. As a result, more people are living with diseases and disability, for longer periods of time (Crimmins and Beltrán-Sánchez 2011; Parker and Thorslund 2007). Therefore, in addition to primary prevention (targeting etiological mechanisms of diseases), secondary and tertiary prevention (reducing disease burden and preventing additional morbidity) have become more important. The challenge is not only to delay the onset of disease, but also to foster socially productive and satisfying lives for older people in poor health.

Social Engagement

Social participation of older people involves both paid and unpaid ways of social contributions. In addition to paid work, it includes volunteering, informal caregiving, learning activities, leisure activities, and religious activities. It has been shown to improve well-being, functioning, and survival among older people. Therefore, we expect that increasing the level of participation particularly in older people with chronic diseases will improve their health and functioning (Glass et al. 1999; Luoh and Herzog 2002; Menec 2003).

The Disablement Process

The definitions of morbidity and disability that were used in this study are based on 'the Disablement Process' (Verbrugge and Jette 1994), which '(1) describes how chronic and acute conditions affect functioning in specific body systems, generic physical and mental actions, and activities of daily life, and (2) describes the personal and environmental factors that speed or slow disablement, namely, risk factors, interventions, and exacerbators.'

The disablement process is a biopsychosocial model (Jette 2006): it integrates both the medical and the social models of disability by acknowledging that disability is influenced by biological, personal, and social forces. Pathology, the clinical manifestation of disease, is linked to impairments and functional limitations, which characterize the dysfunction of body systems and restrictions experienced by people in activities, respectively. Disability, in turn, can be caused by functional limitations, but is also influenced by social circumstances: disability is the gap between personal capability and the demands of the environment. In addition to the main pathway, the model identifies risk factors (predisposing individual-level factors that were present before the onset of the disease), intra-individual factors (behaviour changes, adaptation), and extra-individual factors (medical care, built environment). These factors can buffer or exacerbate the disablement process. Thus, morbidity may, in some cases but not in others, lead to disability.

In this work package, the focus was on morbidity, disability, and social participation. Morbidity is defined as having chronic diseases, which are a combination of pathology and impairments: it includes both diseases that were diagnosed based on clinical measures and diseases that have only overt impairments. According to the disablement process, actually two approaches exist regarding the definition of social participation and disability: the degree to which a person can participate in society can be regarded as an outcome of the disablement process or as a buffering intra-individual factor ('activity accommodations'). These approaches emphasize the reciprocal process of health and participation (i.e. better health leads to more participation, and more participation leads to better health) and both approaches will be taken at certain points in our work.

Approach

In the various phases of our work, the focus was on different parts of the disablement model (Table 6.1). In the first phase, the focus was on trends in HLE in the older European population. Here, we also reviewed trends in the prevalence of chronic morbidity and activity limitations. In the second phase, the focus was on how health status and other individual factors are associated with the level of social engagement of older people with and without health limitations.

Table 6.1 Aims of each phase in the study

Phase 1	Determining trends in HLE and important health indicators
Phase 2	Finding correlates of social engagement in older people with and without health limitations
Phase 3	Fostering social engagement at meso-level
Phase 4	Fostering social engagement at macro-level
Phase 5	Projecting HLE based on social engagement

In the following phases, the focus was no longer on the individual, but on the extra-individual factors. In the disablement model, the extra-individual factors are broadly defined. In this study, a distinction was made between meso-level and macro-level organisations that can contribute to an enabling environment for older people to be socially engaged. Macro-level organisations were defined as governmental organisations and meso-level as non-governmental organisations.

In the final phase, projections were made on the possible effect of social engagement and to what extent it can improve HLE.

Phase 1: Trends in Healthy Life Expectancy

Longer lives in general could imply that people live more years in good health, but also that people live more years with disabilities and, eventually, in a state of dependence. Estimating the number of years that people can expect to live in good health provides crucial information for policy makers. In 2004–2005, LE without activity limitations was added to the EU's structural indicators in the European Union Statistics on Income and Living Conditions (EU-SILC). Therefore, this indicator was used in this study to present recent trends in HLE. Activity limitations were measured using the Global Activity Limitations Indicator (GALI). LE without activity limitations is calculated annually by Eurostat and is also used to set targets for the European active healthy ageing strategy, which aims for a two-year increase in healthy life years on average in EU-27 by 2020 (Robine et al. 2013). Trends in HLE are also compared with overall LE (both healthy and unhealthy years) in the 27 EU countries. Furthermore, trends in the prevalence of two important health indicators are reported: self-reported chronic morbidity and long-term activity limitations.

Life Expectancy With and Without Activity Limitations from 50, 65 and 85 Years of Age

Between 2007 and 2015, LE in the EU27 showed a greater increase in men than in women. Specifically, at age 50 LE increased by 1.3 years from 28.7 to 30.0 years for men and by 0.9 years from 33.7 to 34.6 years for

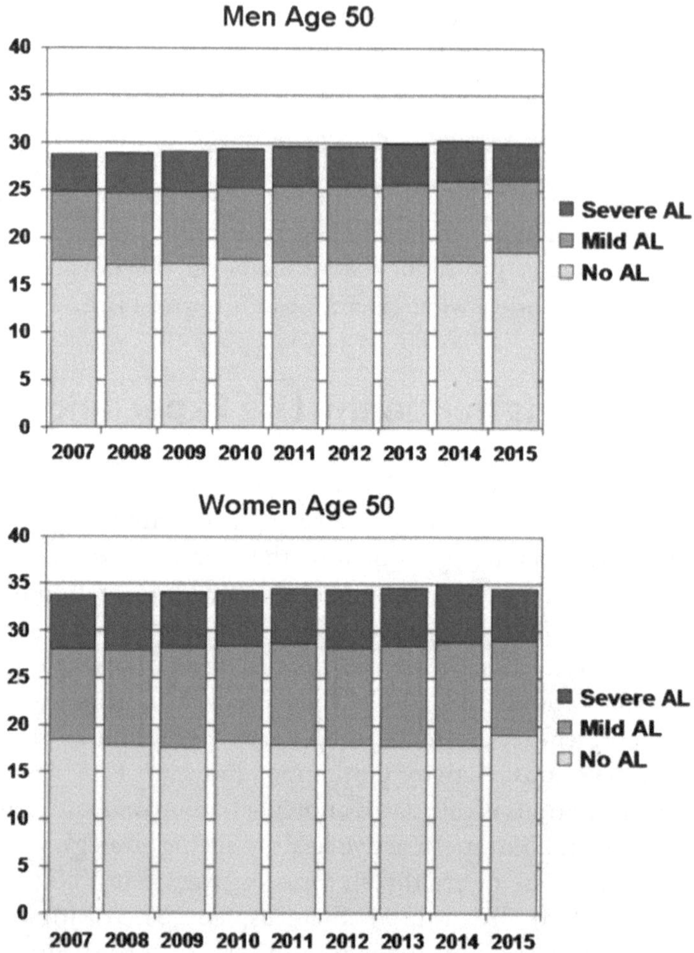

Fig. 6.1 Life expectancy and healthy life years of men and women in the EU-27, at age 50, 2007–2015. (Note: *AL* activity limitation)

women (Fig. 6.1). Meanwhile at age 50, LE without activity limitation (HLE) remained more or less stable. In men this was at 17.6 years, and in women, at 17.8 years. In 2015, however, there seemed to be some improvement in both sexes: in this year, men could expect 18.5 years without activity limitation, and women, 19.0 years. Correspondingly, in 2015 men

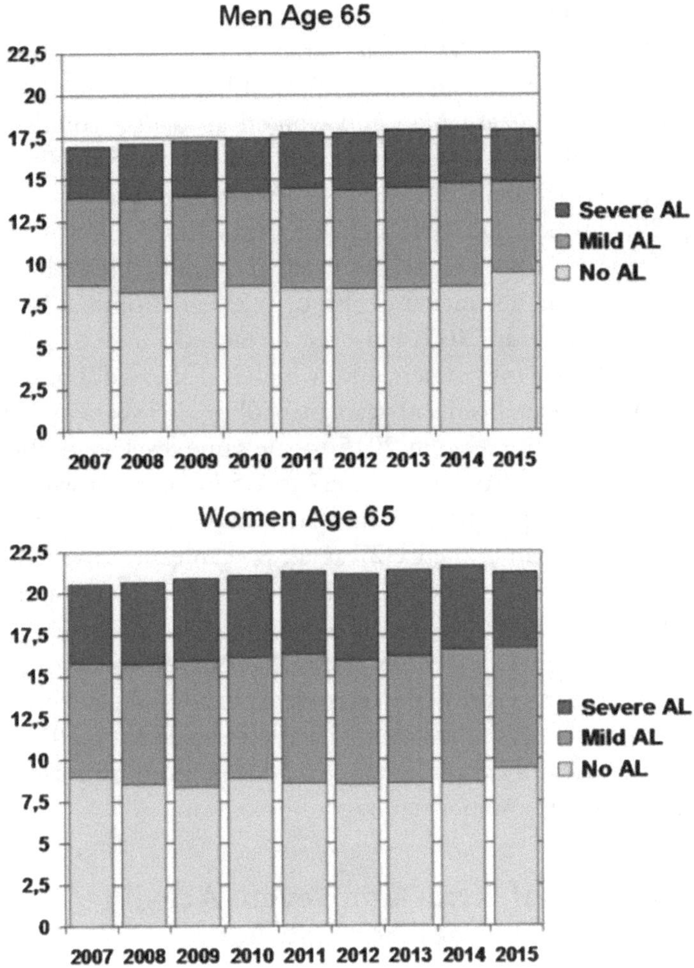

Fig. 6.2 Life expectancy and healthy life years of men and women in the EU-27, at age 65, 2007–2015. (Note: *AL* activity limitation)

and women at age 50 could expect to spend 61.7% and 54.9% respectively of their remaining life without long-term activity limitations.

Turning to LE at age 65 in the EU27, between 2007 and 2015 this increased by 0.9 years from 17.0 to 17.9 years for men and by 0.7 years from 20.5 to 21.2 years for women (Fig. 6.2). Like at age 50, HLE at age

65 remained stable, at about 8.7 years for men and 8.6 years for women, during 2007–2014. In 2015, a remarkable increase was observed. In 2015 in the EU27, men could expect 9.4 HLEs, and women, 9.5 HLEs. Correspondingly, in 2015 men and women at age 65 could expect to spend 52.5% and 44.8% respectively of their remaining life without long-term activity limitations.

At age 85, between 2007 and 2015 LE in the EU27 LE was more variable than for the younger ages. It showed an overall increase by 0.2 years from 5.8 to 6.0 years for men and also by 0.2 years from 6.8 to 7.0 years for women (Fig. 6.3). In 2011, however, it reached a high of 6.2 years for men and of 7.3 years for women, after which it decreased. For men, HLE decreased by 0.4 years from 2.0 years in 2007 to 1.6 years in 2014, after which it showed an increase in 2015 to the same level as in 2007. HLE decreased by 0.5 years from 2.0 in 2007 to 1.5 in 2015 for women, without the improvement in the last year that was observed in men. In 2015, then, men and women in the EU27 at age 85 can expect to spend 33.3% and 21.4% respectively of their remaining life without long-term activity limitations.

In sum, although LE showed clear increases except at age 85, the changes observed in healthy life years were very small. There were year-to-year fluctuations, with the largest change observed from 2014 to 2015, that is, an increase in HLE. It remains to be seen if this improvement will continue or if it was a somewhat larger fluctuation.

The Prevalence of Long-Term Severe Activity Limitations by Five-Year Age Category

The prevalence of severe activity limitations depends on age and sex (Fig. 6.4). The prevalence is higher in older age groups than in younger age groups, varying from 8% (50–54 years) to 41% (85+). The prevalence among women is higher than among men, especially at older ages.

In the EU27, the prevalence of severe limitations in activities among older people (50 years and over) was mostly stable between 2005 and 2015, although after 2013 some decline was apparent in men aged 70–84 as well as in women aged 65–84. However, in men aged 55–59 and in

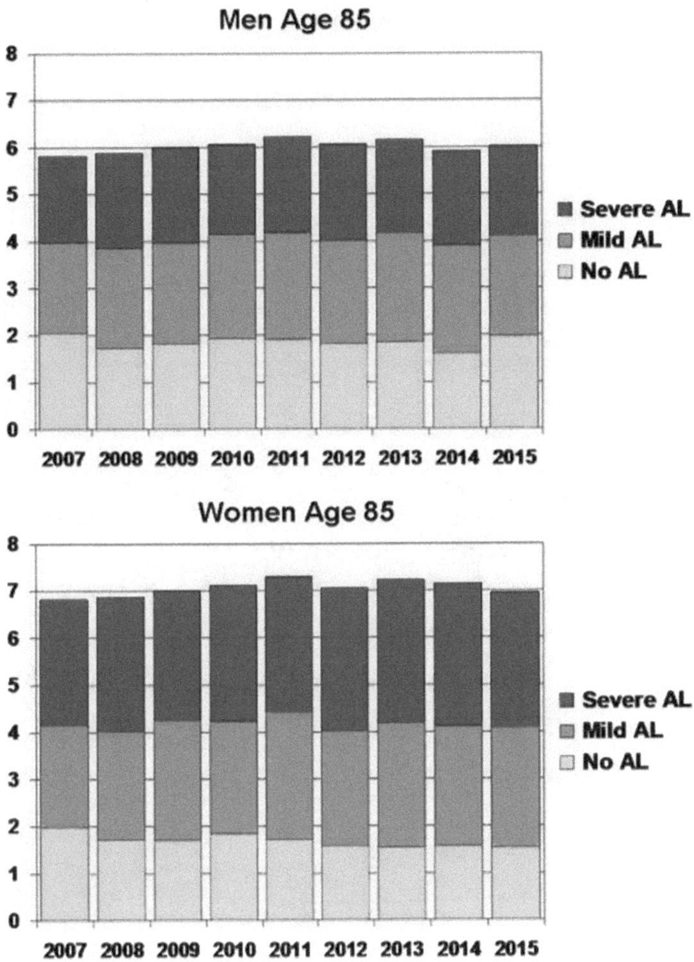

Fig. 6.3 Life expectancy and healthy life years of men and women in the EU-27, at age 85, 2007–2015. (Note: *AL* activity limitation)

women aged 60–64, a small but steady increase was observed. Also, in women in the oldest age group of 85 and over, a steady increase was apparent.

Also the prevalence of mild activity limitations generally showed stability between 2005 and 2015 (Fig. 6.5). However, in men aged 60–64 and 75 and over as well as in women aged 70 and over, small but steady

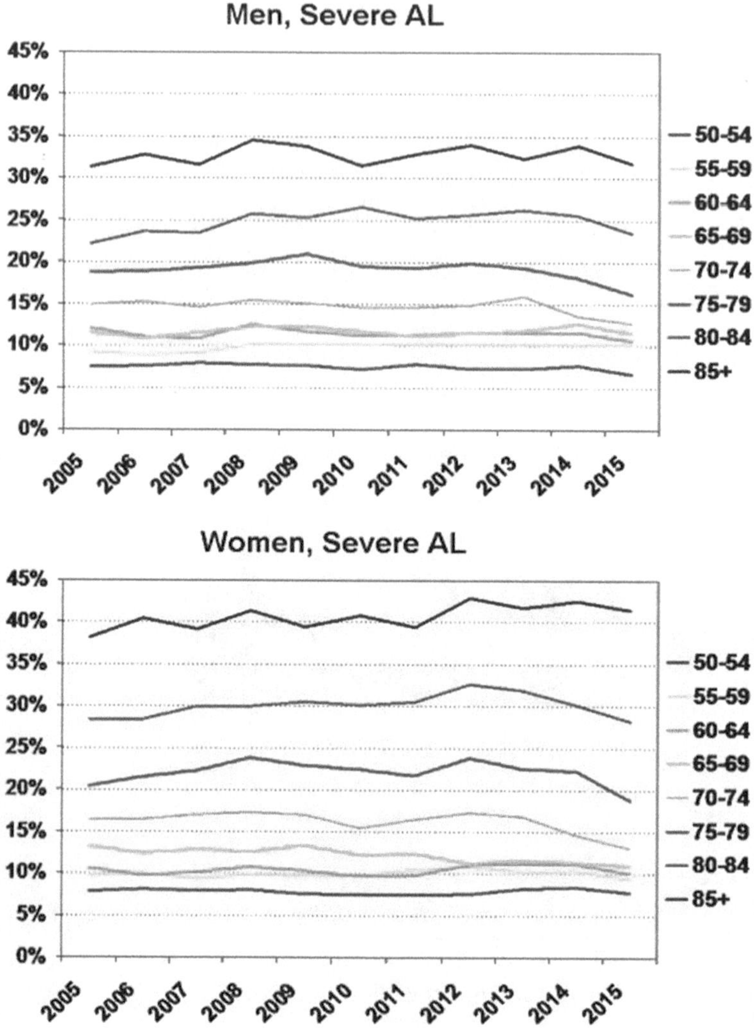

Fig. 6.4 Long-term severe activity limitations in the EU27, in 2005–2015. (Note: *AL* activity limitation)

increases were observed. In all age groups, there was a notable decrease from 2014 to 2015. This decrease corresponds to the decrease in severe activity limitations observed from 2013 in some, but not all age groups. Again, it remains to be seen if this improvement continues.

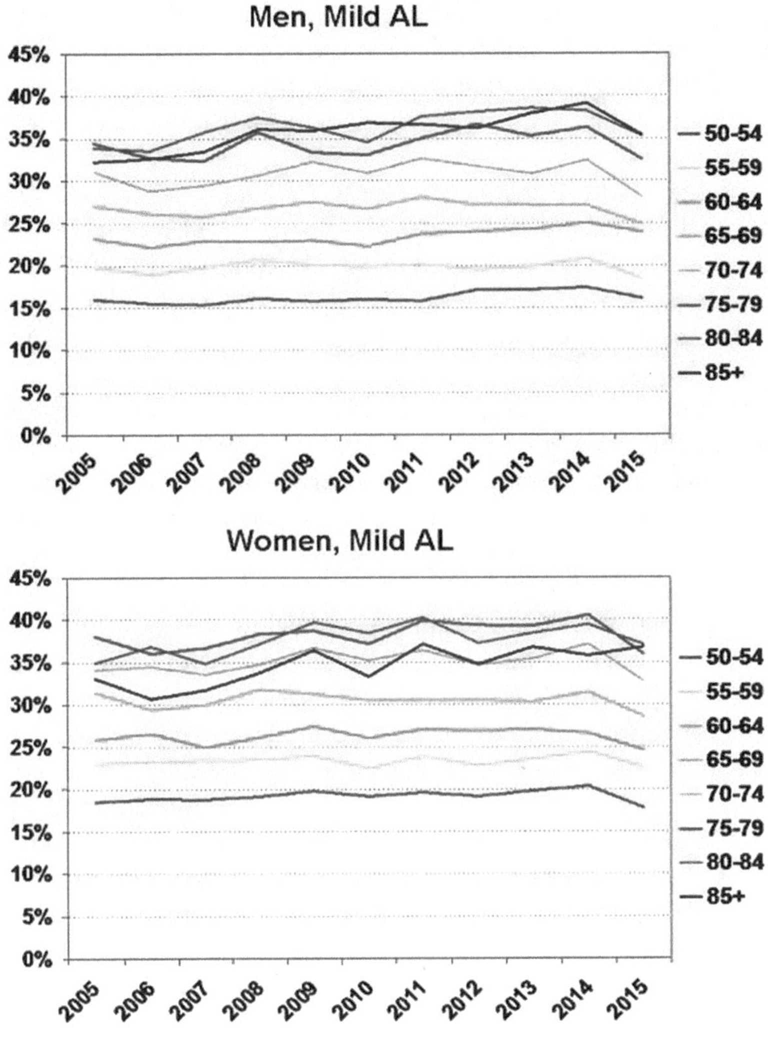

Fig. 6.5 Long-term mild activity limitations in the EU27, in 2005–2015. (Note: *AL* activity limitation)

Conclusions on Healthy Life Expectancy

Across all EU27 countries, LE at age 50, 65, and 85 has significantly increased between 2007 and 2015. This improvement was not accompanied by an improvement in healthy life years: LE without activity limitations for both men and women remained largely stable, as did LE with severe activity limitation. Due to the sudden improvement in mild activity limitations from 2014 to 2015, the often reported increase in LE with mild disability (Christensen et al. 2009) was not observed over the period studied.

On average, the difference between LE and HLE at age 50 was 12.2 years for men and 17.2 years for women. The goal of increasing the HLE with two years on average in all EU countries by 2020 may be reached in particular if functioning is improved in those countries where the HLE is low, while overall LE is relatively high. The four countries with the largest difference between LE and HLE for both men and women were Slovakia, Estonia, Germany, and Italy. In the first three countries, the proportion of years lived in poor health is highest, so improvement in functioning may have large benefits in terms of HLE. In contrast, in Italy the relatively large gap was due to its relatively high LE combined with lower than average HLE. On the other hand, increases in HLE may be gained in all countries where the level of HLE is low, irrespective of their overall LE. Countries with a low level of HLE (less than 15 years for men or women aged 50) combined with a low LE are the Baltic states, Romania, Slovakia, and Hungary.

In most countries, the gap between LE and HLE has increased, because LE has increased more than HLE. Notably, in the EU27 from 2014 to 2015, a narrowing of this gap was observed. However, this was equally attributed to an improvement in HLE and to a decline in LE. As stated earlier, a change over only the last two years cannot be interpreted as a positive development. For this, a longer observation period is needed.

Leading causes for disability among people aged 50 and over are hearing loss, eye disorders, osteoarthritis, ischaemic heart disease, and chronic lung disease. At older ages, stroke and dementia become more important (Rodrigues et al. 2012). Declines in activity limitations can be achieved when for these diseases better treatments are implemented. As these are chronic diseases, treatments will not be able to fully cure them. Therefore, disease management should be a high priority in order to maintain functioning in people living with a chronic disease.

A first limitation of this report is that people in nursing homes are not included in EU-SILC. As a result, the prevalence of activity limitations is underestimated in the figures presented. A decrease in the percentage of older people that live in nursing homes could likely result in an observed increase in the prevalence of activity limitations in the oldest ages. In fact, this is what we observed, but this study does not allow causal interpretations of such observations.

Secondly, although the most complete and up to date data on HLE in the 27 EU countries were used for this report, the time period for which trends could be analysed for all countries together was only nine years. Using a relatively short time period to study trends increases the possibility that the reported trends are chance fluctuations. However, given that time trends in activity limitation were often reflected in the other measures for the majority of countries, the harmonisation of the GALI question is unlikely to be the cause of a change in the distribution of activity limitation by age over the time period (Jagger et al. 2013).

Finally, in several of the countries there has been a change in the wording of the limitation question, which may have influenced the trends in severe activity limitations and in HLE. For example, in 2005 only 3 countries had their version of the GALI question classified as 'fully comparable' to the standard version, 10 countries were classified as 'partially comparable' and 12 as 'not comparable'. In 2009, this was improved to 13 countries being assessed as 'fully comparable', 8 as 'partially comparable and 5 as 'not comparable'. An analysis on the effects of this harmonization across the EU countries has shown that trends in activity limitation between 2005 and 2008 were also reflected in other health measures (Jagger et al. 2013). Moreover, as this report covers at least seven years after the wording changes, these cannot serve as an explanation for the observed trends in HLE and activity limitation.

Phase 2: Determinants of Social Involvement in Older People with Poor Health

Increasing older people's participation is an important goal of current policy, stressed, for example, by the 2012 European Year of Active Ageing (European Commission 2013). Participation is a broad concept, encompassing both activities that have economic value (such as volunteering, labour market participation, and caregiving) and activities that are more consumptive (such as learning activities, religious activities, and leisure activities). Both types of activities can have a strong positive impact on older people's health and well-being.

Older people in better health generally participate more than those in poorer health. Given the current trend of demographic ageing and thus an increasing prevalence of chronic diseases and multimorbidity (the co-existence of two or more chronic diseases), there is a need to develop strategies that promote the social participation of people with chronic diseases in particular. Unfortunately, little evidence exists on the specific determinants of participation in those with poor health.

We address the multidimensionality of active ageing by examining determinants of six different types of social participation—labour market participation, volunteering, caregiving, participation in leisure, participation in educational activities, and church attendance—and focus specifically on differences between groups with and without multimorbidity. Five of the studies use data from SHARE (Survey of Health, Ageing and Retirement in Europe, Börsch-Supan et al. 2005), and one is based on data from EPOSA (European Project on Osteoarthritis, van der Pas et al. 2013).

Determinants of Social Participation in Older People With and Without Multimorbidity

One might expect that people with chronic diseases not only participate less but also that people with chronic diseases experience different barriers that keep them from social participation. For example, older people with multiple chronic diseases may as a consequence suffer from functional limitations affecting their mobility and the ease with which they can participate in activities outside the home. Those with multiple chronic diseases may also need to invest more time in managing their conditions, reducing the time they have available for being socially engaged. Despite the increasing salience of the topic, studies that investigate the level of participation and its predictors, and that focus on people in poor health, are scarce. Therefore, this study focuses on this very topic:

what are the factors that are associated with social engagement in older people with poor health? Several health indicators can be used to distinguish between good and poor health, for instance, self-rated health, physical functioning, or the number of chronic diseases. We operationalise poor health as having multiple health problems, that is, multimorbidity (the occurrence of two or more chronic diseases), or multimorbidity combined with disability. Multimorbidity has proven negative effects on well-being and functioning (Marengoni et al. 2011).

Across most types of participation, findings show considerable differences in the level of participation between those with and without multimorbidity, in favour of the group in better health. Regardless, determinants of social participation generally appear to be quite similar between the multimorbidity and non-multimorbidity groups. Many determinants that are found resemble those of previous studies. In both the multimorbidity and non-multimorbidity groups, and across the different types of participation, factors that are associated with a higher level of participation include a higher socioeconomic status, a larger social network, being active in other types of participation and having better physical and psychological health. If differences are present, they do not always confirm the hypothesis that other resources for social participation compensate for the decline in health in those with multimorbidity: determinants appeared stronger more often in the group without multimorbidity. Depression is recurrently associated with less participation in the group with multimorbidity, indicating that preventing or treating depressive symptoms might be a way to increase the potential of social participation in those with physical health problems. Below we highlight some notable findings that are specific per type of participation.

Principi et al. (2016), examining predictors of volunteering, show that in particular older people with multimorbidity are more likely to be volunteering if they are widowed (rather than married or cohabiting). Previous studies often showed the reverse, but this finding possibly indicates that some people in poor health who lost a relationship compensate for this by volunteering. Principi et al. also show that depressive symptoms are negatively associated with volunteer work particularly for those with multimorbidity or disability.

Labour market participation was studied in the age group 60–69 (Moreira et al. 2016). Both in terms of participation in the labour market and in the number of hours paid work, labour participation by people in poor health is significantly below that of healthy individuals. This is particularly the case in Greece, France, Israel, and the Netherlands, where the difference between healthy and unhealthy individuals is more pronounced. Being involved in sports and social activities is positively correlated with labour market participation, but only for healthy seniors. Among individuals with multimorbidities, higher wages are associated with fewer weekly working hours, and also care responsibilities have a negative effect on the number of working hours.

Determinants of informal caregiving were studied by Schmidt et al. (2016). They conclude that opportunity structures and family structures do not compensate for the effect of multimorbidity on providing care outside the household (to dependent others or to grandchildren). For example, higher education and income, or being married, do not show stronger associations with providing care in those with multimorbidity, compared to those without multimorbidity. With regard to sociodemographic characteristics, they find that women were more likely to provide grandparental care than men, except when they had multimorbidity.

Focusing on educational activities, Golinowska et al. (2016) show that indicators of human capital (e.g. education and income level) are important predictors of participation in formal learning activities. As human capital factors are more difficult to change at older ages than social capital factors, the authors point to the finding that engagement in other social activities might stimulate participation in learning activities.

Participation in social leisure activities proves to be rather high in the six European countries studied by Galenkamp et al. (2016). Interventions may therefore be targeted at the intensity of participation, in particular in those with poor health. People who make use of public transport or of their own car (in particular those with multimorbidity) have higher levels of participation in leisure activities. Thus, improving the transportation possibilities for older people may be a way to remove barriers for leisure participation.

Sowa et al. (2016), finally, studied the determinants of religious participation in older Europeans. This is a type of social participation that people with and without multimorbidity participate in at a rather equal level. Within both groups, people with secondary education—but not with higher education—report lower church attendance compared to primary educated people. This study also finds dementia to be a factor that decreases the chance that people participate in church activities. Other types of social participation show rather strong links with church attendance, in particular in the group without multimorbidity.

All studies to some extent address the fact that activities are not done in isolation, by including other types of participation as predictors. In general, being involved in one type of activity increases the chance of being involved in other activities, providing evidence for the hypothesis that similar motivations are responsible for participating in different types of activities (Morrow-Howell 2010), or that network ties reinforce engagement in other activities (Berkman et al. 2000). On the other hand, a large amount of time allocated to a certain activity can inhibit the participation in other activities. This may apply in particular to labour market participation, a factor that is negatively associated with the outcomes volunteering and grandparenting (Schmidt et al. 2016; Principi et al. 2016). Because patterns of complementing or competing activities are probably better predictors of health and well-being (Morrow-Howell 2010), future studies should address in particular combinations of activities, and their health effects. A related topic concerns the relatively unexplored field of activity preferences. Compared with healthy individuals, do people with health problems take up or give up some activities earlier than others?

A specific determinant that is addressed in the studies is the role of physical functioning within older people with multimorbidity. The disablement process model (Verbrugge and Jette 1994) considers the level of functioning in the pathway between chronic diseases and participation. Including physical functioning as a determinant for both health groups might imply an overadjustment (as it takes away the differences between the multimorbidity and no multimorbidity groups). However, the studies that did adjust for limitations in physical functioning find rather similar results before and after adjustment. This might imply that other health differences between the multimorbidity groups account for some of the determinants that were found to differ between the groups. More likely, the differences between the multimorbidity groups—at least in factors that are associated with participation—are too small to be affected by the level of functioning.

Conclusions on Determinants of Social Participation

The individual studies do provide evidence that health status is related to older people's level of participation in various activities. The challenge is then to enable older people to be active if it is indeed their preference to be socially engaged, despite their health limitations. Nevertheless, based on the results we can also conclude that the level of health should be only a minor factor in designing intervention strategies: little or no substantial differences in predictive factors of social participation were observed. Still, increasing the participation of older people with health problems should remain an important target, not in the least because this might be beneficial for their well-being.

Across types of participation, findings showed that there are considerable differences in the level of participation between those with and without multimorbidity, in favour of the group in better health. However, determinants of social participation appeared to be quite similar between the multimorbidity groups. In both groups, factors that were associated with a higher level of participation were, amongst others, a higher socio-economic status, a larger social network, being active in other types of participation and having better physical and psychological health status. If differences were observed between multimorbidity groups, this was often related to opportunity structures (e.g. indicated by income level or home ownership) being more important for the participation of those without multimorbidity, as compared to the older people with multimorbidity.

Phase 3: Fostering Social Engagement on the Meso-Level

In the previous part of this chapter, we have seen that there are differences in participation rates between older people with and without health limitations. In this part, the focus is on the infrastructure for social participation of older people with health limitations on the meso-level. The goal is to provide insight in the success factors, barriers, and preconditions of current initiatives that foster social participation of older people with health limitations.

In the study we focused on four countries that represent different European welfare regime types, namely the conservative (Austria), the former USSR (Estonia), the post-communist (Poland), and the social-democratic (Netherlands) (Fenger 2007; Warburton and Jeppsson Grassman 2011). We acknowledge that different categorizations are possible (Arts and Gelissen 2002), nonetheless the countries form a meaningful comparison set, reflecting different backgrounds and paths of development.

Moreover, the four countries show quite some variation in participation rates. The Netherlands is among the countries with the highest participation rates, both in people with and without multimorbidity. Austria has somewhat lower participation rates than the Netherlands, but differences between people with and without multimorbidity are smaller. Finally, Poland and Estonia have rather low participation rates. Table 6.2 shows the participation rates in the four countries.

Earlier in this chapter, we have distinguished six dimensions of social participation, reflecting both productive and consumptive types of participation: labour market participation, volunteering, informal caregiving, leisure activities, educational activities, and religious activities. Among the selected countries the levels of participation differ between these various dimensions of participation. For example, in Poland and Estonia labour market participation among older people is high, while other types of participation lag behind, except for informal caregiving. In Austria and the Netherlands, participation rates for the various types of participation are high. Typically, participation rates in volunteering and leisure (not labour market) are higher in northern and western European countries in comparison with eastern and southern Europe (see Fig. 6.6).

In this part of the chapter, the focus is on the infrastructure that fosters social participation of older people with health limitations. In the four different countries, the current infrastructure of meso-level organizations was explored. A limited number of projects fostering social participation was selected for a more in-depth analysis of the success factors, barriers,

Table 6.2 Social participation rates in Austria, Estonia, the Netherlands, and Poland by multimorbidity status (MM)

		Labour market participation SHARE, 50–70		Care for grandchildren SHARE, 50+		Informal care inside the household SHARE, 50+		Leisure activities SHARE, 50+		Volunteering SHARE, 50+		Participation in educational activities SHARE, 50+		Church attendance SHARE, 50+	
	Multimorbidity SHARE, 50+ (%)	No MM (%)	MM (%)	No MM (%)	MM (%)	No MM (%)	MM (%)	No MM (%)	MM (%)	No MM (%)	MM (%)	No MM (%)	MM (%)	No MM (%)	MM (%)
Austria	20.8	37.0	18.3	25.5	21.7	4.6	5.6	29.7	23.3	18.0	17.9	13.9	9.6	19.7	21.4
Estonia	33.4	58.5	32.0	30.5	22.1	5.6	6.8	19.3	12.0	8.4	5.9	16.1	6.1	5.1	5.5
Netherlands	12.5	48.2	24.9	38.3	35.8	5.1	5.2	45.8	39.0	39.4	31.0	21.1	10.6	17.9	16.1
Poland	22.5	27.0	10.4	35.4	31.9	5.9	8.4	4.7	2.4	2.2	2.2	3.4	0.4	41.0	42.3

Source: SHARE data, wave 4

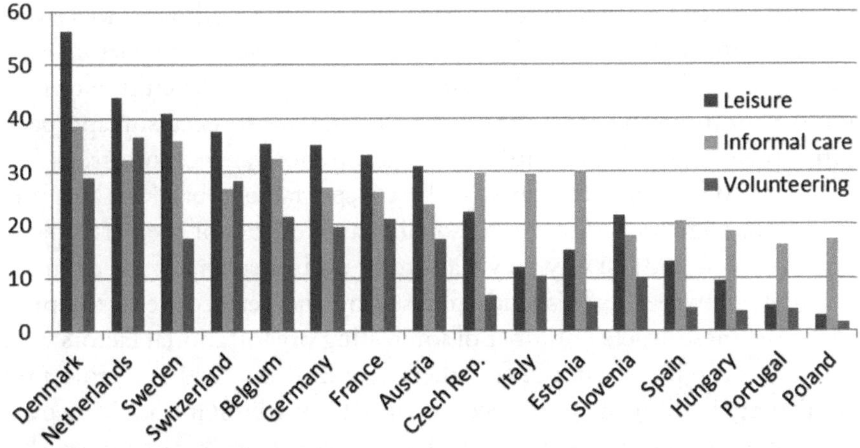

Fig. 6.6 Percentage over people of 65 year and older that report to be involved in leisure activities, informal caregiving and volunteering in selected European countries. (Source: SHARE data, wave 4)

Table 6.3 Overview of dimensions of participation assessed in the selected countries

	Volunteering	Informal caregiving	Leisure activities
Austria		x	
Estonia	X	x	
Poland	X		X
Netherlands	X		X

and preconditions that the projects encounter while engaging older people with health limitations in social participation. In the selected countries, the focus was on different dimensions of social participation. Table 6.3 gives an overview of the dimensions of social participation that were assessed in the different countries.

The projects that were selected aim to engage older people in volunteering, informal care giving, or leisure activities. For example, the Pärnu Two Bridges sports Club[1] in Estonia engages older people in the organization of sports events; the Granny Service[2] in Austria matches older caregivers to families seeking care for children; and the Dutch foundation 'Celebrate life' (Vier het Leven[3]) organizes cultural activities for older people.

Many of the success factors, barriers, and preconditions that were identified in the four countries overlapped. However, some factors were specific for each country as they are closely related to the culture or economic conditions in a country. In that case, the most successful approach differs per country, for example, strategies to motivate volunteers.

As was to be expected many of the success factors, barriers, and preconditions were related to the personal characteristics of the participants in the projects and the way they want to be addressed and involved in the activities. However, as this study focused on the perspective of the projects at the meso-level, a number of interesting organizational factors were found that can be very useful for other organizations starting a project for social participation for older people with health limitations. Moreover, recommendations can be given on how to create an infrastructure that supports the project in reaching their goals and facilitating social participation for older people with health limitations.

In general, there could be more attention for active ageing in societies. The changes in demographics are likely to affect all different levels of society in the future. It is important to be aware of this and, moreover, to react to the challenges that may come. However, it appears that currently society has a negative image of older people and ageing. Older people with health limitations are seen as weak and pitiful. Many organizations are not aware of the fact that older people can be a valuable resource and are therefore not actively recruiting older people. By creating more awareness and sharing knowledge about active ageing more organizations may be stimulated to recruit older people as volunteers or informal caregivers. A change in perspective towards older people may not only be beneficial for participation in organizations, but in general participation may also be stimulated at earlier ages. When people have a habit of being active at a younger age, they are more likely to stay active at older ages (Agahi et al. 2006; Verbrugge et al. 1996).

It was found that financial sustainability was a concern for many of the projects. To create a more supportive infrastructure for the projects, the financial system in the social sector in the different countries should be assessed and changes should be made. Currently, most systems are highly competitive and often exclude social care projects, young organizations, and organizations with a lack of expertise on writing applications. The application procedure for funding from municipalities should be more inclusive towards social care projects. Moreover, there is a demand for more long-term funding or cooperation in order to improve the sustainability of social participation projects.

In Estonia and Poland, older people have less opportunities to participate. Most of them have to work to make ends meet, as a result of insufficient pension benefits. In countries like Austria and the Netherlands, the pension benefits are sufficient and enable older people to participate in social activities other than paid work. People in eastern European countries should have the same opportunities: when people have the desire to participate that should be possible. Therefore, more financial support will be necessary. Pension benefits could be adjusted, or special benefits could be put in place for social participation activities.

In order to improve social participation of older people with and without health limitations, their ICT skills should be improved. This would make it easier for the projects to communicate with participants. More importantly, it would enable older people to be more independent and keep in contact with their own social network online. For example, skype is a widely used technology to keep in contact with relatives. Other innovative examples are online communities like City-towns ('Stadsdorpen') in the Netherlands and AskAround ('FragNebenan') in Austria, where older people can reach out to (older) people in the neighbourhood for help and companionship. Special courses and help-desk can be introduced for digital education of older people.

Phase 4: Fostering Social Engagement on the Macro-Level

The main goal of this part of the chapter is to provide information about macro-level policies that foster health and social involvement of people with health limitations, in selected countries. Given the actual participation rates in each country (Fig. 6.6), a next step is to examine what macro-factors might contribute to the high or low level of social participation in those with multimorbidity, and to the larger or smaller differences compared to those who do not have multimorbidity. We now focus on four countries that show quite some variation in participation rates: Austria, Italy, Poland, and the Netherlands. The Netherlands is among the countries with the highest participation rates, both in people with and without multimorbidity. Austria and Italy have somewhat lower participation rates of people with multimorbidity than the Netherlands, but differences between people with and without multimorbidity are smaller. Finally, Poland has rather low participation rates (an exception is participation in religious activities).

These four countries also represent four European welfare regime types, namely the conservative (Austria), the Southern European (Italy), the post-communist (Poland), and the social-democratic (Netherlands). Again, we acknowledge that different categorizations are also possible, but the countries nevertheless form a meaningful comparison set, reflecting different backgrounds and paths of development. This concerns, for example, the role of families in welfare provision and the availability and level of disability and pension benefits.

Room for improvement seems to exist in the level of social participation of older people with and without health limitations. Some countries have clearly higher participation rates than others, and overall, people without health problems participate more than people with health problems. However, we cannot conclude too quickly that the participation rates in all countries could be raised to those in the best performing country. A few cautionary remarks should be made. Historical, cultural, and policy contexts differ across countries, so it remains unsure if social participation rates in some countries can actually be improved to reach those of countries with the highest participation rates. The extent to which current or future European policies may contribute to these improvements is difficult to conclude from our study. Recent changes in participation are not always the result of or triggered by the introduced policies, but, for example, may result from attitudinal influences that are hard to change. Demographic changes, changes in population health, and contradicting policies may have their effects on the rate of social participation as well.

Recent studies on health trends of older people reveal that the prevalence of chronic diseases is currently increasing (Crimmins and Beltrán-Sánchez 2011; Parker and Thorslund 2007). The difference in participation rates between people with and without multimorbidity suggests that the future overall level of participation may be negatively affected by this higher disease prevalence. In this light, it is important to note that in each of the countries only few policies exist that are targeted at increasing the social participation of older people with poor health. The negative effect of poor health status might be counteracted by other developments, such as higher educational level and policy measures. In addition, prevention programmes, for example, those promoting healthy lifestyles are needed both at young and older ages, to indirectly increase social participation (Broese van Groenou and Deeg 2010). A nice Austrian example of a preventive programme is fit2work, which aims to prevent morbidity and disability and improve employability at all ages (www.fit2work.at).

Older people have only limited time available for their preferred activities or commitments. While (multiple) roles may have mutual positive effects between them (Burr et al. 2005), competing roles may also prevent people from taking up new activities, for example, due to lack of time. Policies to change the participation in one domain (e.g. labour market participation) may thus at the same time have effects on other domains of participation, such as caregiving and volunteering. Additional regulations have the potential to manage this effect, for example, those that exist for combining paid work and caregiving responsibilities.

Compared with policies and regulations to increase leisure participation or volunteering, labour market policy reforms have been quite radical, but for a large part have only recently been put in place. Their effects on the social participation of working age seniors—other than that they increase the share of older people who are in paid work—are largely unknown.

A final cautionary remark concerns decentralization and the role of regional and local governments in stimulating social participation of older people. In Austria and Italy, many policies and services do not exist at the national level, but only at the regional level. Decentralization with regard to social care has taken place only very recently in the Netherlands and Poland. In general, this leads to larger diversity and it remains unknown to what extent this improves the conditions for older people in all regions, or only in some.

Across the four countries, policies that most explicitly address the social participation of older people concern labour market policies. Pension reforms, that have primarily been put in place to offset budgetary constraints, have contributed to the increase in labour market participation of many older people. This increase might also result from younger cohorts having less physically demanding jobs, that enable them to continue working longer. Nevertheless, the labour market is a field where (recent) regulations are directed specifically towards people with disabilities, at least in Austria, the Netherlands, and Italy. Although these might be improved, for example, in Italy, they are expected to lead to an increase in the work participation of this group in particular. At the same time, the demographic trend of an increasing number of older people in need of support and care asks for increasing participation in volunteering and informal caregiving. To deal with the challenges posed by the demographic change, including the consequences for health care and pension costs, and to avoid competition between different roles, one recommendation is to develop policies to enable working people to meet the care demands they experience. In some countries such regulations are in place already, and the challenge will be to make those attractive both for employers and employees.

To a lesser extent, policies and regulations exist to improve social participation in two other 'productive' participation domains: caregiving and volunteering. However, policy changes in the long-term care domain may in some countries lead to a larger call for volunteers and informal carers. As this often concerns people with health limitations themselves, it is remarkable that services in these domains are often not targeted at older people or those with disabilities. For example, informal caregiving within the household—often to a spouse—can be quite burdensome and recommendations involve the availability of appropriate respite care for informal carers. This enables them to continue their provision of care for a longer period and may reduce the negative psychological and health effects they may encounter.

With regard to volunteering, in some countries volunteer organizations support and train volunteers and match supply and demand for certain types of volunteer work. It was noted that such organizations need sufficient professional coordinators. Thus, funds for hiring such coordinators need to be at a sufficient level. In addition, specific recruitment programmes and incentives are needed to include older people—with or without health limitations—in volunteer work.

Leisure activities that are more consumptive, such as entertaining social contacts, learning and sports activities, are performed in the personal sphere, and as such have the least interference from policies. Nevertheless, specific programmes or benefits may be introduced that foster the leisure participation of people with poor health in particular. One social participation domain that is worth mentioning is lifelong learning: an increasing number of older people take part in Third Age Universities in Poland, Italy, and the Netherlands. Investing in learning from a life-course perspective in current societies also means that learning, working, and social participation are not bound to specific age groups anymore. For example, focusing on social participation—other than work—of younger aged groups may be beneficial for them and for society, because this increases the chance that they continue their activities when they age.

Phase 5: Projections of Healthy Life Expectancy

As described in the beginning of this chapter, social participation is regarded as a buffer in the disablement process, considering that participation ('activity accommodations') may prevent further functional decline (e.g. Menec 2003). In this part of the chapter, we examine the extent of disability in people with mulitimorbidity who participate in social activities. We consider the promotion of social participation among older individuals with multimorbidity as one tool to increase disability-free life expectancy (DFLE or HLE). Hence, we expect that raising the level of social participation will reduce the prevalence of disability, and thus increase life years without disability.

For three scenarios that foster social engagement, the effects on HLE are assessed for four selected countries. Scenarios are based on changes in social participation in the older population over ten years, and estimates of the effect of social participation (volunteering, informal caregiving, labour participation, educational activities, leisure activities, religious activities) on disability, derived from the empirical literature and own estimates. For the latter, as for earlier analyses reported in the chapter, we use data from the SHARE, wave 4.

Three scenarios of increases in social participation that seem achievable within the coming ten years were developed and studied.

- The No differences scenario: the first scenario is based on the idea that barriers within a country that prevent people with multimorbidity from achieving the same level of social participation as their countrymen without multimorbidity can be removed. As a result, the level of social participation of older people with multimorbidity is raised to the level of those without multimorbidity within each country.
- The Maximum scenario: in the second scenario, we assume that countries can learn from each other and that institutional and policy changes can raise the level of social participation among individuals with multimorbidity to the level of the best performing country.
- The Cautious scenario: the third scenario assumes that the maximum attainable participation rate for people with multimorbidity is given by the level of the best performing country. However, social and institutional differences between countries make it impossible to raise the level of participation of individuals with multimorbidity beyond the current level for individuals without multimorbidity within each country.

Given the fact that increases in social participation are assumed to take place within a relatively short time period of ten years, the maximum increase in all scenarios was limited to 10%-points. Even then, the scenarios are quite ambitious. Different dimensions of social participation were not distinguished; they are all assumed to rise. We raise the level of each form of social participation at the same time. Possible time constraints were ignored: raising the participation rate in one domain (e.g. formal work) is likely to affect the time available for other dimensions of participation (e.g. informal caregiving); however, this was not taken into account.

Against our expectation, our findings show that all three scenarios of increasing social participation rates of older individuals with multimorbidity have a very modest effect on DFLE. Our scenarios show improvements of 0.0–0.1 years. Even scenarios with high increases in social participation rates and high estimates of the effects of social participation on disability result in improvements in HLE of around 0.2 years. Our findings suggest that substantial increases in HLE cannot be expected to result from a single intervention alone, but will have to be achieved by combined efforts on several fronts.

However, preventing disability is not the only reason why policymakers should focus on improving social participation of older individuals. Being socially active at older ages can be important in its own right. Social participation could have positive effects on other aspects of well-being besides disability, that were not included in this analysis. Given the ageing of the population, and the increasing prevalence of chronic diseases, focusing on social participation among older individuals with multimorbidity is of special importance.

Implications

Between 2007 and 2015, LE at age 50, 65, and 85 has significantly increased across all European countries. However, LE without activity limitations for both men and women showed no increase and instead, general stability when considering the EU27 overall. In most countries, the gap between LE and HLE is increasing.

People in better health generally participate more in social activities than those in poorer health. Therefore, people in poor health in particular should be stimulated to be socially active. However, the level of health of the target group should be only a minor factor in designing intervention strategies: little or no substantial differences in predictive factors of social participation were observed between people with and without health limitations.

Social participation rates may be raised by establishing a stimulating infrastructure on meso- and macro-level. The best practices for meso-level organisations to involve older people with health limitations in social activities differ between countries, and the economic and cultural context should always be taken into account when attempting to transfer best practices from one country to another.

Most macro-level policies that stimulate social participation of older people are focused on labour market participation. This interferes with the participation in other dimensions. These other dimensions, such as volunteering and informal caregiving, are also important, so policies should also focus on them. Moreover, governmental policies could support meso-level organisations to complement the offer of social activities for older people.

When the social participation rates improve due to changes in policy, it is not likely that this will improve HLE. However, social participation may still be beneficial for the well-being of older people as well as for society at large. Thus, efforts to improve the infrastructure for social activities are valuable.

Research Priorities

While our work has produced new insights, it also points to a number of priorities for further research. In the first place, monitoring of trends in morbidity, disability, and HLE should be continued across European countries. Second, as extending working lives is high on the political agenda, it is important to pay particular attention to other social activities that may compete in time with labour market participation. In particular, this pertains to other productive activities including volunteering and caregiving. Related to this point, more insight needs to be gained on what older workers need, so that they can continue non-paid social activities. Third, we expected social participation to be an important factor in improving health, in particular in people with multimorbidity. However, this expectation was not supported by our data. A further research priority is to study a variety of factors or combination of factors that have the potential to increase healthy life years. If enough of such factors are found, HLE may increase at the same rate—or faster—as total LE. In turn, the potential of active ageing may be realised.

Notes

1. http://www.2silda.ee/defaulten
2. http://www.familie.at/site/wien/angebote/omadienst
3. https://4hetleven.nl/

References

Agahi, N., Ahacic, K., & Parker, M. G. (2006). Continuity of Leisure Participation from Middle Age to Old Age. *The Journals of Gerontology Series B: Psychological Sciences and Social Sciences, 61*(6), S340–S346. https://doi.org/10.1093/geronb/61.6.S340.

Arts, W., & Gelissen, J. (2002). Three Worlds of Welfare Capitalism or More? A State-of-the-Art Report. *Journal of European Social Policy, 12*(2), 137–158.

Berkman, L. F., Glass, T., Brissette, I., & Seeman, T. E. (2000). From Social Integration to Health: Durkheim in the New Millennium. *Social Science & Medicine, 51*(6), 843–857. https://doi.org/10.1016/S0277-9536(00)00065-4.

Börsch-Supan, A., Hank, K., & Jürges, H. (2005). A New Comprehensive and International View on Ageing: Introducing the 'Survey of Health, Ageing and Retirement in Europe'. *European Journal of Ageing, 2*(4), 245–253. https://doi.org/10.1007/s10433-005-0014-9.

Broese van Groenou, M. I., & Deeg, D. J. (2010). Formal and Informal Social Participation of the 'Young-Old' in The Netherlands in 1992 and 2002. *Ageing and Society, 30*(03), 445–465. https://doi.org/10.1017/S0144686X09990638.

Burr, J. A., Choi, N. G., Mutchler, J. E., & Caro, F. G. (2005). Caregiving and Volunteering: Are Private and Public Helping Behaviors Linked? *The Journals of Gerontology Series B: Psychological Sciences and Social Sciences, 60*(5), 247–256.

Christensen, K., Doblhammer, G., Rau, R., & Vaupel, J. W. (2009). Ageing Populations: The Challenges Ahead. *The Lancet, 374*(9696), 1196–1208. https://doi.org/10.1016/S0140-6736(09)61460-4.

Crimmins, E. M., & Beltrán-Sánchez, H. (2011). Mortality and Morbidity Trends: Is There Compression of Morbidity? *The Journals of Gerontology Series B: Psychological Sciences and Social Sciences, 66B*(1), 75–86.

Deeg, D. J. H., van Vliet, M. J. G., Kardaun, J. W. P. F., & Huisman, M. (2013). Understanding the Mortality Decline at Older Age. Improved Life Course or Improved Present Period? *Annual Review of Gerontology and Geriatrics, 33*, 261–291.

European Commission. (2013). *About the European Innovation Partnership on Active and Healthy Ageing, Innovation Union – A Europe 2020 initiative.* Retrieved December 9, 2014, from http://ec.europa.eu/research/innovation-union/index_en.cfm?section=active-healthy-ageing&pg=about

Fenger, M. (2007). Welfare Regimes in Central and Eastern Europe: Incorporating Post-Communist Countries in a Welfare Regime Typology. *Contemporary Issues and Ideas in Social Sciences, 3*(2), 1–30.

Galenkamp, H., Gagliardi, C., Principi, A., Golinowska, S., Moreira, A., Schmidt, A. E., Winkelmann, J., Sowa, A., van der Pas, S., & Deeg, D. J. (2016). Predictors of Social Leisure Activities in Older Europeans With and Without Multimorbidity. *European Journal of Ageing, 13*(2), 129–143. https://doi.org/10.1007/s10433-016-0375-2.

Golinowska, S., Sowa, A., Deeg, D., Socci, M., Principi, A., Rodrigues, R., Ilinca, S., & Galenkamp, H. (2016). Participation in Formal Learning Activities of Older Europeans in Poor and Good Health. *European Journal of Ageing, 13*(2), 115–127. https://doi.org/10.1007/s10433-016-0371-6.

Glass, T. A., Mendes de Leon, C., Marottoli, R. A., & Berkman, L. F. (1999). Population Based Study of Social and Productive Activities as Predictors of Survival Among Elderly Americans. *British Medical Journal, 319*(7208), 478–483.

Jagger, C., Fouweather, T., & EHLEIS Team. (2013). *Cross-Sectional Analysis of Health Expectancies in 2008: Evaluation of the 2008 Implementation of the Greater Harmonisation of the Mini European Health Module.* EHLEIS Technical_report_2013_5.1, www.ehemu.eu

Jette, A. M. (2006). Toward a Common Language for Function, Disability, and Health. *Physical Therapy, 86*(5), 726–734.

Luoh, M. C., & Herzog, A. R. (2002). Individual Consequences of Volunteer and Paid Work in Old Age: Health and Mortality. *Journal of Health and Social Behavior, 43*(4), 490–509. https://doi.org/10.2307/3090239.

Marengoni, A., Angleman, S., Melis, R., Mangialasche, F., Karp, A., Garmen, A., Meinow, B., & Fratiglioni, L. (2011). Aging with Multimorbidity: A Systematic Review of the Literature. *Ageing Research Reviews, 10*(4), 430–439. https://doi.org/10.1016/j.arr.2011.03.003.

Menec, V. H. (2003). The Relation Between Everyday Activities and Successful Aging: A 6-Year Longitudinal Study. *The Journals of Gerontology Series B: Psychological Sciences and Social Sciences, 58*(2), S74–S82.

Moreira, A., Antunes, C., Ilinca, S., Rodrigues, R., Principi, A., Sowa, A., Galenkamp, H., & Deeg, D. (2016). *Poor Health and the Labour Supply of Senior Workers* (Unpublished Manuscript).

Morrow-Howell, N. (2010). Volunteering in Later Life: Research Frontiers. *The Journals of Gerontology Series B: Psychological Sciences and Social Sciences, 65*(4), 461–469.

Parker, M. G., & Thorslund, M. (2007). Health Trends in the Elderly Population: Getting Better and Getting Worse. *The Gerontologist, 47*(2), 150–158. https://doi.org/10.1093/geront/47.2.150.

Principi, A., Galenkamp, H., Papa, R., Socci, M., Suanet, B., Schmidt, A., et al. (2016). Do Predictors of Volunteering in Older Age Differ by Health Status? *European Journal of Ageing, 13*(2), 91–102. https://doi.org/10.1007/s10433-016-0377-0.

Robine, J. M., Cambois, E., Nusselder, W., Jeune, B., Van Oyen, H., & Jagger, C. (2013). The Joint Action on Healthy Life Years (JA: EHLEIS). *Archives of Public Health, 71*(1), 2.

Rodrigues, R., Huber, M., & Lamura, G. (2012). *Facts and Figures on Healthy Ageing and Long-Term Care*. Vienna: European Centre for Social Welfare Policy and Research.

Schmidt, A. E., Ilinca, S., Schulmann, K., Rodrigues, R., Principi, A., Barbabella, F., et al. (2016). Fit for Caring: Factors Associated with Informal Care Provision by Older Caregivers With and Without Multimorbidity. *European Journal of Ageing, 13*(2), 103–113. https://doi.org/10.1007/s10433-016-0373-4.

SHARE, Survey of Health, Ageing and Retirement. www.share-project.org

Sowa, A., Golinowska, S., Deeg, D., Principi, A., Casanova, G., Schulmann, K., Ilinca, S., Rodriques, R., Moreira, A., & Galenkamp, H. (2016). Predictors of Religious Participation of Older Europeans in Good and Poor Health. *European Journal of Ageing, 13*(2), 145–157. https://doi.org/10.1007/s10433-016-0367-2.

van der Pas, S., Castell, M. V., Cooper, C., Denkinger, M., Dennison, E. M., Edwards, M. H., Herbolsheimer, F., Limongi, F., Lips, P., Maggi, S., Nasell, H., Nikolaus, T., Otero, A., Pedersen, N. L., Peter, R., Sanchez-Martinez, M., Schaap, L. A., Zambon, S., van Schoor, N. M., & Deeg, D. J. H. (2013). European Project on Osteoarthritis: Design of a Six-Cohort Study on the Personal and Societal Burden of Osteoarthritis in an Older European Population. *BMC Musculoskeletal Disorders, 14*(1), 138.

Verbrugge, L. M., & Jette, A. M. (1994). The Disablement Process. *Social Science & Medicine, 38*(1), 1–14. https://doi.org/10.1016/0277-9536(94)90294-1.

Verbrugge, L. M., Gruber-Baldini, A. L., & Fozard, J. L. (1996). Age Differences and Age Changes in Activities: Baltimore Longitudinal Study of Aging. *The*

Journals of Gerontology Series B: Psychological Sciences and Social Sciences, 51(1), S30–S41.

Warburton, J., & Jeppsson Grassman, E. (2011). Variations in Older People's Social and Productive Ageing Activities Across Different Social Welfare Regimes. *International Journal of Social Welfare, 20*(2), 180–191. https://doi.org/10.1111/j.1468-2397.2009.00691.x.

7

Improving Health in Later Life: How a Life Course Approach Could Improve Health and Well-Being in Old Age

Carmen Giefing-Kröll and Beatrix Grubeck-Loebenstein

Introduction

We are experiencing a dramatic demographic shift: in Europe, the share of people 65 years and older will increase from 18.9% in 2015 (according to Eurostat statistics) to 29.3% in 2060 (Lanzieri 2011). Due to the increase in life expectancy, we are facing medical, social, and economic challenges that arise from population ageing. Improvement in healthy life years (HLY), that is, disability-free life expectancy (LE) is one of the main health goals for the European Union (EU). However, as highlighted in the previous chapter, no increase in HLY was observed over the past decade (Fig. 7.1). Novel strategies are needed.

Body functions decline with increasing age and we become more vulnerable to age-related diseases: central obesity, metabolic syndrome,

C. Giefing-Kröll • B. Grubeck-Loebenstein (✉)
Institute for Biomedical Aging Research, University of Innsbruck, Innsbruck, Austria
e-mail: Carmen.giefing-kroell@uibk.ac.at; Beatrix.Grubeck-Loebenstein@uibk.ac.at; https://www.uibk.ac.at/iba; https://www.uibk.ac.at/iba

A. Walker (ed.), *The Future of Ageing in Europe*,
https://doi.org/10.1007/978-981-13-1417-9_7

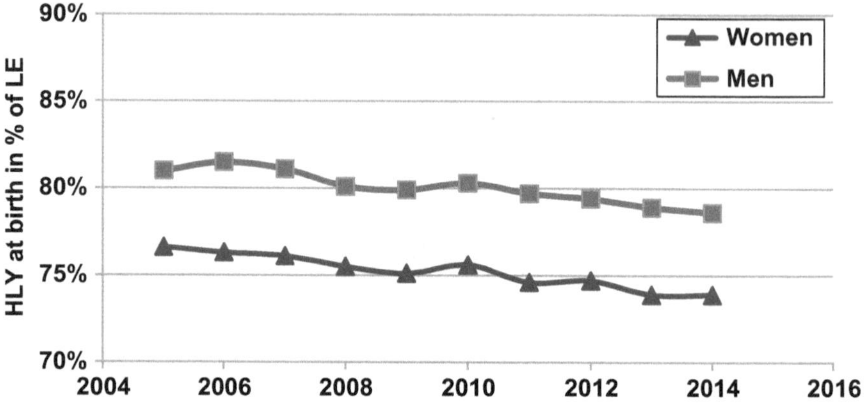

Fig. 7.1 Healthy life years (HLY) at birth in per cent of life expectancy (LE). (Data source: Eurostat, European Core Health Indicators data tool. http://ec.europa.eu/health/dyna/echi/datatool)

neurodegeneration, cancer, asthma, depression, diabetes, osteoporosis, sarcopenia, and cardiovascular disease (de Cabo et al. 2014).

However, large differences in ageing can be observed between individuals: one of the most striking examples is the longer LE of women compared to men whereas spending more years in ill health. Women live longer than men in general, and also the record for the longest living human is held by the French woman Jeanne Calment with an age of 122.5 years. In addition, centenarians and supercentenarians, that is, people living beyond 100 or 110 years of age, respectively, are living examples of successful ageing. Most centenarians are remarkably healthy up to old age and experience a rapid decline late in life, resulting in compression of morbidity in their final years (Caruso et al. 2012). Thus, we know that it is biologically possible to reach extreme longevity and spend most of it in good health. Biogerontology tries to find the answer how to achieve this.

We need a fundamental understanding of the biological mechanisms that sustain life and health in order to develop suitable intervention strategies to prolong healthy life and prevent morbidity and disability. Consequently, the key questions biogerontology tries to answer are: which mechanisms are responsible for the accumulation of damages with age, how do biological and lifestyle factors influence healthy LE, and which intervention strategies are suitable to improve lifelong health and well-being.

We know now that ageing is not a chaotic process but determined by genes and biochemical processes. The interaction of genes and the environment determines our rate of ageing. Nevertheless, we are not the slaves of our genes: longitudinal studies of monozygous twins have shown that only about 20% of the variance in longevity is due to inherited genes. Environmental factors are far more important (Steves et al. 2012). Studies on long-lived individuals show that 25% of age of death in adulthood is heritable; this increases with advancing age (Brooks-Wilson 2013). Consequently, biogerontology tries to discover which intervention strategies throughout lifetime are suitable to maximise HLY.

The Key Challenges

- LE is increasing
- No increase in HLY, that is, years lived without disability over the past decade despite increase in LE
- A thorough understanding of the mechanisms that are responsible for age-associated changes in cells and organs is needed
- This knowledge enables the development of suitable intervention strategies to improve lifelong health and well-being
- Translation of evidence-based intervention strategies requires efficient communication with relevant stakeholders to ensure broad application in the population

State-of-the-Art Biogerontology from Womb to Tomb

Biogerontology is important at every stage of life. In order to successfully extend HLY, we need to adopt a life-course perspective in healthy ageing. Interventions must be applied early enough to prevent physical degeneration and decline in cognitive function. A life-course perspective on ageing and risk of developing age-related diseases is essential: plasticity and thus the possibility to intervene decreases with age, whereas damage increases. Interventions in adults, especially those at risk, are beneficial to a certain degree, also screening in middle-aged reduces the risk for non-communicable diseases; however, interventions in adolescents and young adults could be more effective and also reduce the disease risk in the next generations.

The developmental period is highly plastic in order to respond to environmental conditions and allow fine-tuning of gene expression through epigenetic regulation in an evolutionary sense to improve the reproductive capacity and sustain health (Feil and Fraga 2012). This is especially true for the prenatal and early postnatal phase during tissue development but extends to childhood and puberty for some tissues such as the brain and reproductive system. During this developmental period, the organism is highly sensitive to environmental factors such as nutrients, environmental chemicals, drugs, and infections. Interference with this epigenetic regulation is implicated in disease development later in life including major public health problems: obesity, diabetes, hypertension, cardiovascular disease, asthma and allergy, immune and autoimmune diseases, neurodevelopmental and neurodegenerative diseases, precocious puberty, infertility, some cancer types, osteoporosis, depression, schizophrenia, and sarcopenia (Barouki et al. 2012).

In addition to lifestyle interventions such as fasting regimes, caloric restriction, or exercise, a limited number of pharmacological and genetic interventions have sound evidence to promote healthy lifespan including spermidine, resveratrol, metformin, and rapamycin (de Cabo et al. 2014).

Based on biogerontological and epidemiological findings, we elucidate in the following sub-sections which interventions at which stage of life are needed to preserve health and prevent or ameliorate disease.

Foetal Life

The foundation for successful ageing is already laid in the womb. Retrospective studies of children of the Dutch Famine 1944/45 have shown that there is a strong correlation between malnutrition during pregnancy and increased risk for non-communicable disease development in adult life (Gluckman et al. 2008). Disease risk is increased, if there is a mismatch between the expected — for example, malnutrition of the foetus — and the actual living environment — for example, overnutrition (Gluckman et al. 2008). The relation between prenatal nutrition and later risk for metabolic disease seems to be U-shaped: overnutrition of the foetus increases the risk of the child to suffer from obesity, type 2 diabetes, cardiovascular disease, and cancer later in life (Darnton-Hill et al. 2004; Wakeford and Bithell 2015; Goodell and Rando 2015). Nutrition during foetal life seems to be imprinted on the epigenome and transferred to the next generation in maternal and paternal line without affecting the genome (Feil and Fraga 2012; Pembrey et al. 2014).

Not only a poor maternal diet but also exposure to pollutants in the air, soil, and food during foetal development and the first years of life affects the development of the immune system and increases the risk for non-communicable diseases such as obesity, diabetes, cardiovascular disease, asthma, allergies, or infertility in adult life (Hanson and Gluckman 2014; Barouki et al. 2012; Debost-Legrand et al. 2016; Feil and Fraga 2012; MacGillivray and Kollmann 2014). In addition, symptomatic but also asymptomatic maternal infections affect our healthy ageing potential maybe through alterations in the developing immune system (Mazumder et al. 2010; Vaiserman 2011; MacGillivray and Kollmann 2014).

Studies indicate that the induced phenotypes might be reversible by dietary intervention or pharmacological manipulation (Gluckman et al. 2008). In summary, a balanced diet, prevention of infectious diseases through vaccination, and limiting the exposure to pollutants already during pregnancy constitute the basis of healthy ageing for the growing human being. Politics need to improve nutrition and reduce exposure to environmental chemicals before and during pregnancy and in the first years of life. We need to invest in the education of young people in relation to their responsibility during pregnancy and parenthood to foster health of subsequent generations.

Infancy and Childhood

The way we are born seems already to affect our potential for healthy ageing through its influence on the intestinal flora. Children delivered via Caesarean have an increased risk for obesity, celiac disease, asthma, and diabetes (Yuan et al. 2016; Paliy et al. 2014).

Nutrition continues to be a crucial factor to delay or foster the onset of certain diseases. Infants that are exclusively breastfed during their first four – six months of life are 40–70% less likely to catch an infection and show 50% less hospital admissions during the first year of life compared to formula fed infants (Duijts et al. 2010; Dewey et al. 1995; Forsyth 1995; Paricio Talayero et al. 2006). Current data indicate that breastfed infants have a lower risk for obesity, type 2 diabetes, and other parameters of the metabolic syndrome during adult life compared to formula fed infants (Hypponen et al. 1999; von Kries et al. 1999; Toschke et al. 2002; Paliy et al. 2014). Data regarding allergy prevention is conflicting, overall no clear protective effect of breastfeeding on the development of allergies such as asthma, atopic dermatitis, or eczema could be shown (Bergmann et al. 2002; Kull et al. 2005; Kramer et al. 2007). Bioactive components in breast milk induce epigenetic changes with possible implications on health later in life (Ruchat et al. 2014). The neonatal nutritional environment may play an important role in the development of obesity through long-term programming effects on feeding behaviour, satiety signalling, and metabolism (Hale et al. 2015). Compared to the evidence from breastfeeding, the impact of differences in weaning, that is, the gradual introduction of solid food and its implications on long-term health are less well understood (Robinson 2015).

A growing body of evidence connects health problems in adulthood and old age with infections and health conditions early in life (Bengtsson and Lindström 2003; Yi et al. 2007; Monteverde et al. 2009). Consequently, vaccination plays an important role from the start. Even if the individual does not develop serious symptoms upon infection, the struggle against the pathogenic invader leads to increased levels of inflammation that contribute to an increased risk for cardiovascular disease later in life.

Persistent organic pollutants or phospho-organic pesticides — also known as environmental obesogens — may play role in the current obesity epidemics by altering the glucose metabolism (Debost-Legrand et al. 2016). Especially the role of bisphenol A is repeatedly investigated in this context (Carwile and Michels 2011; Vafeiadi et al. 2016).

Physical activity and sedentary behaviour habits formed during childhood tend to remain through adulthood. To combat current obesity epidemics and inactivity crisis regular physical activity for children should be encouraged through active outdoor play (Tremblay et al. 2015). Together with genetics and nutrition, physical exercise during childhood and adolescence also determines peak bone mass that influences the ultimate risk of osteoporotic fracture later in life (Michel et al. 2008).

To what extent interventions early in life affect healthy ageing still needs further investigation. Studies with twins indicate that less than 10% of LE is determined by conditions during infancy and childhood (Steves et al. 2012). However, studies in animal models show that satiety, food, and exercise capacity are determined by developmental and biological factors. Once set points are established in early life, these are difficult to reverse explaining the difficulty in sustaining lifestyle interventions (Hanson and Gluckman 2014).

Adolescence

At this transition to adulthood, risk behaviours are initiated: over 90% of lifetime tobacco alcohol and drug use is initiated in this phase of life. More than three quarters of mental health problems, obesity, asthma, and autoimmune conditions are present before reaching adulthood (Viner et al. 2011). During adolescence, risk factors for later diseases develop, for example, high blood pressure that is strongly related to the risk of cardiovascular disease or stroke later in life (McCarron et al. 2000). Once these risk factors are established, they tend to track throughout life, for example, overweight and obesity in particular are difficult to combat once they are established (Darnton-Hill et al. 2004).

Adolescents have to a lesser extent benefited from health improvements over the past 50 years than other age groups (Viner et al. 2011).

New interventions are needed to promote health literacy, a healthy lifestyle, and physical exercise in the adolescent population. Health-related behaviour started in adolescence impacts on health across the life course.

Adulthood

Ageing starts early in life: the physical fitness of a 30-year old is already less than the one of a 20-year old (de Cabo et al. 2014). After the age of around 40 years, the deterioration of physical systems becomes detectable, for example, difficulties with memory and learning due to cognitive decline, weakening of skeletal muscles, or declines in bone mineral density (McPhee et al. 2016). Menopause at around 50 years of age has a strong impact on women's health and in particular also on their immune defence mechanisms (Al-Azzawi and Palacios 2009; Giefing-Kroll et al. 2015). Biogerontological research shows that there exist several common features of ageing including genomic instability, telomere attrition, epigenetic alterations, loss of proteostasis, deregulated nutrient-sensing, mitochondrial dysfunction, cellular senescence, stem cell exhaustion, and altered intercellular communication (Partridge 2014). The incidence of various alterations increases with age within a given tissue. However, the range of changes differs between tissues and organs and particularly between individuals (Partridge 2010). In contrast to these intrinsic alterations, other visible age-related changes such as greying of hair are bystander effects with little functional significance.

Adult life may be subdivided into three different stages of prevention: (1) postpone age-related changes in the cell and organism itself, (2) prevent the onset of specific age-related diseases, and (3) ameliorate the living conditions after disease onset by specific disease treatment (Seals and Melov 2014). Animal studies indicate that even if applied in midlife interventions still have the potential to delay the development of age-related diseases substantially (Bitto et al. 2016). There is a strong association between mid-life behaviour, successful ageing, and late-life health (Lafortune et al. 2016). There is strong evidence for immediately applicable lifestyle interventions that are successful to promote healthy ageing and prevent the onset of disease and disability. In addition, the potential of and evidence for novel pharmaceutical substances and other methods to intervene with the ageing process will be discussed in this section. There is a lot of research in diverse fields for potential anti-ageing interventions with mixed proof for efficacy.

Lifestyle Anti-ageing Interventions

A healthy lifestyle is the best-established strategy for healthy ageing. This includes a balanced diet, exercise, regular vaccination, and stress reduction. Stress alters the production of hormones and major life stressors even affect the telomere length (Puterman et al. 2015). Other well-known strategies include a save environment at home and around; reduction in harmful behaviour such as smoking, excessive alcohol consumption, and physical inactivity; and high-quality medical care. The short- and long-term benefits of quitting smoking are well documented. Most importantly, age is no barrier for the beneficial effects of changes in lifestyle: even a 60 years old one can expect to gain three years of LE (Doll et al. 2004). In the following, we discuss the effects of diet, exercise, and vaccination on healthy ageing in more detail.

Dietary Interventions

Our diet has a significant impact on our LE and especially on our health in old age. The best established and most studied intervention strategy is caloric restriction with significant evidence for its impact on healthy ageing in several species from yeast to mice to non-human primates.

Caloric Restriction Caloric restriction tries to achieve health-promoting and life-extending effects by reducing the caloric input but avoiding malnutrition regarding essential nutrients. Whereas in mice this corresponds to a reduction by 30% of calories per day, the amount of restriction in humans has not yet been clearly defined. As a reference point 15% of restriction or not gaining weight after reaching adulthood with the exception of pregnancy was suggested (Willcox and Willcox 2014). Especially people in industrialised countries gain weight with rising age. Caloric restriction reduces the release of growth factors such as growth hormone or insulin, and thus extends the lifespan of yeast, flies, worms, rodents, and dogs (Fontana et al. 2010). Studies with model organisms such as mice or rhesus macaques have shown that a reduction in caloric intake reduces the rate of type 2 diabetes, cardiovascular disease, and cancer by metabolic changes (Fontana and Partridge 2015). In humans, long-term caloric restriction seems also to be protective against type 2 diabetes, high blood pressure, and arteriosclerosis (Fontana 2009). Empirical evidence for its efficacy in humans includes the Biosphere 2 study where human volunteers experienced an unpredicted caloric restriction for 24 months (Walford et al. 2002) or the population of Okinawa, that had a long-term lower caloric input but also a higher life expectancy than the rest of Japan (Willcox and Willcox 2014).

However, other studies have shown that the effect of caloric restriction is not universal but also dependent on genetic background (Mulvey et al. 2014). There are intraspecies variations within different genotypes of mice with some being non-responsive at all to caloric restriction (Sohal and Forster 2014). Especially those mouse strains with a low increase in bodyweight over adulthood in the ad-libidum-fed control group showed no effect questioning the suitability of the control groups, as these obese animals die prematurely (Sohal and Forster 2014).

While most studies in animal models are performed as early-onset caloric restriction diet, it would not be reasonable to apply the same study design in humans (Anderson et al. 2009). In young humans, caloric restriction may negatively impact on development, the attainment of maturity or provoke cognitive defects. Potential negative effects of caloric restriction are related to malnutrition including neurologic deficits, immunodeficiency, amenorrhea, decreased fecundity, impaired wound healing, and osteoporosis (de Cabo et al. 2014; Fontana and Partridge 2015). Animal studies in mice and primates show that adult-onset caloric restriction is also effective in delaying ageing and the onset of age-related diseases (Spindler 2005; Colman et al. 2014). However, this kind of diet is not easy to sustain and requires a high amount of self-discipline to achieve sufficient supply with all nutrients. Consequently, other dietary and pharmacological interventions are investigated to achieve similar effects.

Diet Composition Data from animal studies provide more and more evidence that not only the reduction in calories, but in particular the composition of the diet has a significant impact on the development of age-related diseases. However, the determination of the perfect amount, composition and ratio of just the macronutrients — carbohydrates, lipids, and proteins — is laborious, the possible combinations seem infinite (Fontana and Partridge 2015). In addition, age has an impact on the optimal nutrient composition of the diet. Several popular diets and medical recommendations propagate the consumption of high amounts of protein especially from animal sources to reduce obesity, sarcopenia, osteoporosis, frailty, and surgical stress. In contrast to these recommendations, scientific studies indicate that the reduction in the protein content of a diet or — even more specific — of single amino acids at a constant caloric input has beneficial effects on health and longevity. Especially, the reduction in the amino acid methionine has positive effects (Fontana and Partridge 2015). In addition, animal models show that replacing animal with plant protein reduces tumour growth (Fontana et al. 2013).

Fasting From an evolutionary perspective, only very recently we have unlimited access to food, whereas in earlier times availability was restricted. In addition, fasting is part of most human populations and cultures. This may reflect mammalian adaptation to overeating in periods of food abundance, for example, after successful hunt and starvation in between (de Cabo et al. 2014). As psychologically more viable alternatives to caloric restriction, several fasting regimes are investigated: alternate day fasting (Johnson et al. 2007; Varady et al. 2009; Heilbronn et al. 2005), 5:2 intermittent fasting diet (Harvie et al. 2011), and 48 hours fast once or twice a month (de Cabo et al. 2014). Intermitted fasting that is characterised by alternating periods of food intake and fasting leads to an increase in LE and a decrease in age-associated diseases including neurodegeneration, cancer, and cardiovascular disease in animal models (Longo and Mattson 2014). Fasting in mice increases the efficacy and reduces adverse effects of chemotherapy against certain tumours (Raffaghello et al. 2008). Also in humans, positive effects can be observed (Fontana and Partridge 2015). In addition, the duration of the fasting period has an important impact: if the intake of the same amount of calories is limited to five to seven hours a day compared to five meals a day, this has a positive impact on health (Mattson et al. 2014). Also, if the majority of calories is consumed earlier during the day, this increases weight reduction and has further positive effects by following the natural circadian rhythm (Fontana and Partridge 2015). Epidemiological data show that continuous interruption of the nutrition cycle and the circadian rhythm, for example, shift work, increases the risk of obesity, type 2 diabetes, cardiovascular disease, cancer, and neurodegenerative disorders (Wang et al. 2011). For optimal efficacy, fasting regimens should be combined with a healthy diet such as Mediterranean or Okinawa low-protein diet (de Cabo et al. 2014).

Also, the gut **microbiome** plays an important role. Changes in the diet composition regarding, for example, protein or fibre content alter the microbiome's structure, function, and secreted factors. Also, an imbalance of the microbiome increases the risk for cardiovascular disease, metabolic disorders, cancer, autoimmune, and allergic reactions (Fontana and Partridge 2015).

As described in the previous section, already the nutrition of the mother has significant effects on the later disease risk. Even **transgenerational effects** of nutrition have been described, including the paternal line (Pembrey et al. 2014).

Physical Activity

An active lifestyle with regular physical exercise plays a major role in extending health span. Regular exercise reduces morbidity and mortality in humans and animal models (de Cabo et al. 2014). Physical training and exercise is the second important pillar for the prevention of cardiovascular disease, type 2 diabetes, osteoporosis, sarcopenia, and depression and prolongs independent living of the elderly (de Cabo et al. 2014). The intensity is crucial: burning 2000 kcal through exercise per week promotes health, more than 3000 kcal accelerates the ageing process, for example, marathon runners have increased cortisol levels, and professional athletes rarely live beyond their 80th birthday (Lee and Skerrett

2001). Moderate physical activity, for example, daily 30-minute walks have positive effects on overweight individuals also without weight loss (Laaksonen et al. 2005). Regular resistance exercise reduces the risk of developing osteoporosis and fracture later in life (Ferrucci et al. 2016). An active lifestyle and regular physical exercise has a significant impact on brain health and prevents/reduces depression, anxiety disorders, stress, and the risk of mental degeneration and Alzheimer and improves self-confidence (Groot et al. 2016; Kaliman et al. 2011). Exercise also positively affects several mechanisms involved in ageing including telomerase activity, antioxidant pathway, and mitochondrial biogenesis (Kaliman et al. 2011). In contrast to dietary intervention, regular exercise does not increase LE in animal models, but preserves health in the long run (Mercken et al. 2012). A combination of dietary interventions with physical exercise thus has additive effects (Mercken et al. 2013).

Vaccination

Vaccination is another important healthy ageing tool by primarily preventing infectious diseases, but secondly also by reducing inflammation and thus the risk for the development of age-related diseases. In addition to its positive effects for the individual, on the population level regular vaccination plays an important role in herd immunity, that is, preventing spread of disease in those that cannot be vaccinated being too young, too old or immunocompromised through other means (Baracco et al. 2015; Gavazzi 2009; Prato et al. 2010; Dolan et al. 2012). Regular booster vaccinations during adult life are essential to provide protection in old age when immunosenescence precludes efficient protection. This has been perfectly illustrated for the tetanus vaccine: men who received additional vaccination during military service were better protected against tetanus in old age (Gergen et al. 1995). In order to increase coverage rates in adults a life course vaccination schedule based on age rather than risk is necessary (Poland et al. 2009). In addition, health literacy needs to be improved: people need to be informed on both the benefit and risk of vaccines not only for themselves but also their personal environment, for example, children. Other measures to increase vaccination rates include recommendations through health care professionals, regular checks of vaccination status, and especially the reduction in out-of-pocket costs for vaccines.

Pharmacological Anti-ageing Interventions in Development

The major problem with most lifestyle interventions is the low compliance to follow guidelines for a longer period. Dietary restriction and exercise are both well studied interventions that delay ageing and enhance health span. Due to poor compliance both are difficult to translate to the population level, although the positive effects of a healthy diet and regular physical activity are well known in the general public. Very few follow these recommendations, as short-term satisfaction seems more desirable than the expected long-term positive effects on health. Consequently, the discovery and development of pharmacological interventions are pursued to achieve similar health-preserving effects, for example, caloric restriction mimetics or exercise pills that synergise with lower levels of exercise (Kaeberlein et al. 2015). In order to combat ageing, interventions need to activate the main cellular processes involved in ageing including autophagy, improved stress resistance, and mitochondrial efficiency (de Cabo et al. 2014). The National Institute of Aging (NIA) has a special programme for systematic testing of candidate age-delay treatments. A handful of interventions have convincingly and reproducibly shown to prolong the lives of various strains of mice, for example, a drug called rapamycin (Miller et al. 2011).

Rapamycin This molecule from a soil bacterium is mainly used as immunosuppressant during organ transplantation to reduce rejection. Rapamycin increases lifespan in all model organisms tested including yeast, flies, worms, and mice through inhibition of mTOR (mechanistic target of rapamycin) (Bjedov and Partridge 2011). Rapamycin retards multiple aspects of ageing in mice (Wilkinson et al. 2012) and treatment late in life is sufficient to extend lifespan, reverse cardiac decline, and improve immune function (Johnson et al. 2013). However, due to its strong immune-suppressive effect it is not suitable as anti-ageing drug in humans. Adverse effects include impaired wound healing, anaemia, proteinuria, pneumonitis or hypercholesterolemia (de Cabo et al. 2014). Alternative administration strategies to limit negative effects on health (Anisimov et al. 2011b) and analogues are investigated. In a recent study, an analogue of rapamycin successfully boosted the immune function in elderly (Mannick et al. 2014).

Resveratrol is a polyphenol that is found in grapes and red wine. This candidate molecule became popular due to the so-called French-Paradox: French winemakers suffer less frequently from cardiovascular disease despite a fatty diet. In mouse models, resveratrol reduces the oxidative stress in the heart, neurodegeneration, vascular disease, and type 2 diabetes and increases LE of mice on a high-fat diet (Baur et al. 2006). Also in monkeys on a diet high in fat and sugar resveratrol had positive effects on health (Fiori et al. 2013; Jimenez-Gomez et al. 2013; Mattison et al. 2014). Resveratrol also prolongs the lifespans of flies, worms, and yeast (Baur and Sinclair 2006). Side effects if administered in high doses include discomfort and nausea in humans and nephrotoxicity in mice (Patel et al. 2011). Resveratrol especially protects against stress- and age-associated diseases by acting on several stress-related targets including SIRT1, a member of the longevity-associated protein family of sirtuins (Baur and Sinclair 2006). Sirtuins are also a target for the development of other anti-ageing drugs. Regarding the antioxidant and anti-inflammatory effects of resveratrol, promising data from in vitro experiments and animal models exist, but results in humans are mixed (Conti et al. 2016). Resveratrol improves the insulin sensitivity and glucose control in diabetic but not in non-diabetic patients (Liu et al. 2014). If combined with exercise training, in comparison to training alone resveratrol blunts the positive effects on cardiovascular health (Gliemann et al. 2013). Clinical trials have so far largely focused on the pharmacodynamics and metabolism of resveratrol, future studies will have to show its clinical efficacy (Patel et al. 2011).

Metformin is the most prescribed drug for the treatment of type 2 diabetes. It decreases hepatic gluconeogenesis and increases insulin sensitivity (Berstein 2012). Metformin inhibits chronic inflammation and reduces oxidative stress. The effects of metformin mimic the transcriptional changes induced by caloric restriction (Anisimov et al. 2011a; Martin-Montalvo et al. 2013). Metformin improves health and prolongs the lifespan of mice (De Haes et al. 2014). In a retrospective analysis of observational data, diabetic patients taking metformin had reduced mortality compared to a control group treated without metformin. They also lived longer than matched individuals without diabetes not receiving metformin (Bannister et al. 2014). The question whether metformin is capable to delay ageing in humans will be analysed in the TAME study (Barzilai et al. 2016).

Other interventions that are in development are:

- **Exercise pills.** Rarely patients comply with prescriptions for physical exercise due to lack of time, motivation, and other reasons. The molecular changes that are induced by exercise are now better understood and consequently methods for replication and intervention are developed (Li and Laher 2015).
- **Spermidine** increases LE in yeast, flies, and worms (Eisenberg et al. 2009) and inhibits neurodegeneration in mice (Wang et al. 2012). In humans the concentration of endogenous spermidine decreases with age, but not in centenarians (Pucciarelli et al. 2012).
- Cellular **nicotinamide adenine dinucleotide (NAD^+)** concentrations change during ageing and decreased NAD^+ levels contribute to the ageing process and the development of age-related diseases (Verdin 2015).
- **Senescent cells** accumulate during ageing and secrete factors that promote inflammation and cancer (Campisi and Robert 2014). Three different strategies might be feasible in order to mitigate the deleterious effects of cellular senescence (Tchkonia et al. 2013): (1) Interfering with pathways leading to senescence or delay the development of senescent cells, for example, via caloric restriction. (2) Eliminating senescent cells with specific markers. The removal of senescent cells was shown to slow down the development of pathologies (Baker et al. 2011). (3) Interfering with the senescence-associated secretory phenotype.
- **Circulating factors and stem cells.** Heterochronic parabiosis and transplantation experiments in which an aged animal is joined with a young one via their circulatory system or tissue is transplanted showed

that humoral factors and the local and systemic environment affect age-dependent declines in several tissues, for example, the brain, muscles, the liver, and the heart (Goodell and Rando 2015; Conboy et al. 2015). Young stem cells exposed to an aged milieu showed accelerated ageing whereas placing aged cells in a young environment had positive effects (Bitto and Kaeberlein 2014). A study will test whether blood donated from young adults and infused into patients with mild to moderate Alzheimer's disease can improve their symptoms (Scudellari 2015).

- **Telomeres.** Chronological age accounts for less than 10% of human telomere length variation. Telomere maintenance is a trade-off between maintenance and the development of certain often lethal cancers. Psychological stress might be one factor in telomere attrition in humans (Blackburn et al. 2015).
- **Mitochondrial-targeted interventions.** The relationship between mitochondria and ageing is extremely more complex than the free radical theory of ageing originally suggested (Wang and Hekimi 2015). Mitochondria wear down with ageing, but age-dependent mitochondrial dysfunction is not sufficient to limit lifespan. However, deviations from normal mitochondrial function may extend longevity (Wang and Hekimi 2015; Gonzalez-Freire et al. 2015). In order to boost mitochondrial function several strategies have been suggested including mitochondrial antioxidants and mitochondrial protective peptides (Dai et al. 2014).

Popular Anti-ageing Interventions with Questionable Evidence

Antioxidants As the use of vitamin and mineral supplements is quite popular, there is also a need to provide the scientific evidence regarding benefits beyond the complementation of dietary deficiencies. Studies on supplementation with antioxidants fail to provide evidence for improvement over a well-balanced diet; no impact on general mortality or cancer incidence was observed but partially even detrimental effects (Bjelakovic et al. 2012; Conti et al. 2016; Dolara et al. 2012). Neither epidemiological nor intervention studies could show an effect of megadoses of vitamin C in protecting against atherosclerotic disease or cancer (Hercberg et al.

1998), but resulted even in an increased oxidative DNA damage (Podmore et al. 1998). Beta-carotene and vitamin E seem to increase mortality, as well as higher doses of vitamin A (Bjelakovic et al. 2012). Post-menopausal overdosage of vitamin A increases the risk for hypocalcaemia and hip fractures (Bjelakovic et al. 2012). Supplementation of vitamin E (α-tocopherol) increased the risk of heart failure in several studies (Conti et al. 2016). Overall none of all the tested antioxidants could show any positive anti-ageing effects above the recommended daily intake in any clinical study.

Hormones Due to the changes in hormone production (including sex-steroids, growth hormone, and insulin-like growth factor 1) with increasing age, the administration of hormones as anti-ageing substances is propagated. Currently it is not clear if the medication with hormones has health-improving, life-extending or even life-shorting effects (Zouboulis and Makrantonaki 2012). The most studied example is oestrogen: although unpleasant effects of the menopause and the reduction in bone density can be ameliorated, oestrogen administration is possibly linked with an increased risk of death from cardiovascular disease. The risks and benefits of hormone supplementation in ageing remain largely controversial. Thus, the reduction in endocrine production should not be compensated with the administration of synthetic hormones but with a lifestyle that stimulates the natural endocrine production.

In addition, different kinds of food have received attention for their anti-ageing potential, for example, olive oil, green tea, soy, grapes/red wine, broccoli, curcuma, and garlic. However, the scientific proof for these is limited (Akazawa et al. 2012; Brondino and Re 2014; Oliveras-Lopez et al. 2013; Fleischauer and Arab 2001; Ackermann et al. 2001). Also, the use of herbal remedies is increasing. At the same time herbal therapies are not free of adverse effects and some even interact with prescribed medicine. Data support the efficacy of some popular herbal medicines but considerable knowledge is still lacking to perform reliable risk-benefit assessments (Ernst 2002). Also, for Coenzyme Q10 it is impossible to draw a conclusion on its beneficial or harmful effects on age-related diseases, as studies lack relevant information (Conti et al. 2016).

Elderly: What Can Still Be Done

Although conditions early in life significantly affect human health and disease risk late in life, demographic data indicate that they are less important than changes in current conditions. Also, elderly themselves can largely influence their life and healthy LE via a healthy lifestyle. The acceleration in physical decline caused by external factors is generally reversible at any age. The "German unification" demonstrated that mortality is even highly plastic at advanced ages: after the unification of East and West Germany the mortality levels in the East declined rapidly to Western levels, especially among the elderly (Vaupel et al. 2003). Even in their 80s and 90s people benefit from medical interventions such as cataract surgery or hip replacement. In addition, physical training also increases muscle strength and mobility even among the oldest old and also frail (Vaupel et al. 2003).

In contrast to the current approach following the prevention of specific age-related diseases, an alternative strategy might achieve the compression of morbidity into a short period late in life: by slowing the process of ageing thus delaying the upstream effects of ageing such as inflammation, oxidative stress, mitochondrial dysfunction, cellular senescence, stem cell exhaustion, epigenetic alterations, and genomic instability (Seals and Melov 2014).

The major biological problems elderly are confronted with are cognitive decline, depression, limitations in mobility due to frailty or sarcopenia, cardiovascular disease, malnutrition, and immunosenescence. Health promotion and illness prevention programmes for elderly should include regular influenza and pneumococcal vaccination, physical activity, balanced nutrition, smoking cessation, appropriate treatment of pain, management of osteopenia, detection and treatment of hypertension and diabetes (Morley and Flaherty 2002).

Nutrition In the elderly the main focus in diet is the maintenance of a healthy body weight through balanced nutrition and an appropriate fluid intake. Age-related changes in the gut microbiota and resorption may lead to malnutrition in a range of different micronutrients, for example, iron, magnesium, folic acid, vitamin C and D, in turn leading to a poor immune response, wound healing, or hospital admission (O'Toole and Jeffery 2015). The diet may be supplemented with vitamin D and calcium against osteoporosis. Studies show that vitamin D supplementation is efficient in preventing 10–15% of fractures, in reducing falls and overall mortality by approximately 5% (Allan et al. 2016; Bjelakovic et al. 2014). However, massive doses of vitamin D also have negative effects.

Exercise Physical activity improves respiratory and muscles function, bone health as well as body functioning as a whole and reduces the risk of non-communicable disease, depression and mental degradation. Increased aerobic exercise in the elderly improves blood pressure, lipids, glucose tolerance, bone density, depression and the overall quality of life (Fleg 2012). Regular exercise is a key intervention to prevent falls with injuries which are to a significant extent responsible for disease and disability and a main risk factor for the development of frailty (Todd and Skelton 2004). By reducing inflammation and delaying sarcopenia and disability, regular physical activity is a potential intervention to achieve prolonged independent living (Nicklas and Brinkley 2009). For elderly physical activity includes leisure activities, mobility like walking or cycling, occupational activity, household chores, games, sports or activity during daily familial or social activities (WHO Regional Office for Europe 2004). Resistance exercise in particular has beneficial effects, as it does not only improve strength but also enhances cognition and decreases depression (Fisher and Morley 2002). However, excessive exercise increases mortality in the elderly population.

Hormones Regarding the age-related decline in the levels of enzymes, hormones and other metabolites the general belief is that loss of them is harmful and hormone levels should be set back to a youthful state. However, the observed changes could also be a sign of remodelling and adaptation for survival and health, as reductions in hormones and their receptors have been found to be a requirement for longevity. For example, the long-lived naked mole rat has reduced hormone levels and increased maintenance and repair systems (Rattan 2014). Levels of many hormones decline with age, but presently it is unknown if their supplementation will improve health and increase lifespan or decrease it (Fisher and Morley 2002). Oestrogen can prevent unpleasant menopausal conditions and counteract the rapid loss of bone mineral density but at the same time oestrogen supplementation increases the mortality from cardiovascular disease. Studies on growth hormone have failed to support its role as an anti-ageing hormone, but supplementation had many unpleasant effects (Bartke et al. 2001; Johannsson et al. 2000; von Werder 1999).

Major New Findings Within Mopact

Prevention

Prevention prior to the development of age-associated diseases is the key for successful ageing. Vaccination is a well-established strategy to prevent infectious diseases that account for one-third of mortality in persons above 65 years and are a major factor in the development of morbidity and disability. The efficacy of the immune system deteriorates with age including its defence mechanisms and vaccination success. In addition, there exist biological and socio-economic inequalities in infectious disease susceptibility that contribute to differences in HLY and LE between men and women and across EU member states.

Empirical studies show that there is a clear bias in susceptibility to infectious diseases based on sex and gender. In line with that also responses to some vaccines differ between men and women. To comply with sex-specific differences in immunity, composition, and dosage of vaccines need to be adjusted specifically for men and women in order to increase their efficacy and reduce adverse reactions.

Not only vaccinations in the elderly are required to protect them from disease, but a life-course perspective on vaccination with regular booster immunisations throughout adulthood is needed to sustain protection till old age. Vaccine policies are not consistent across EU member states and often specific recommendations for adult vaccination are missing. The lack of harmonisation of vaccine recommendations and financial compensation programmes across Europe leads to significant differences in vaccination rates and consequently protected individuals. In addition, the public needs to be better educated on the risks and benefits of vaccination to increase compliance with vaccination guidelines. These issues need to be addressed to provide more equal opportunities for healthy ageing.

Intervention

The great majority of age-related chronic diseases can be prevented or delayed with the implementation of healthy nutrition, physical exercise, and cognitive training interventions. Based on current data, dietary manipulations are the most promising interventions to increase healthy LE. Data from animal and human studies indicate that long-term caloric restriction combined with an adequate intake of the required nutrients induces multiple metabolic adaptations that decrease the risk of developing type 2 diabetes, hypertension, cardiovascular disease, and cancer. Recent discoveries indicate that timing of food intake and fasting is also critical: both intermittent fasting and a food intake adjusted to the diurnal rhythm impact on metabolic health and multiple biological functions, even in the absence of modifications in overall calorie intake. Also, the diet composition, for example, lowered intake of proteins or specific amino acids, influences health and longevity in animal models. In addition, nutritional modulation of the gut microbiome seems to affect metabolic pathways that control cardiovascular health, cancer risk, and inflammatory/immune function. More studies are warranted to understand the intricate interactions of factors regulating ageing and age-associated chronic disease. Due to the low compliance to dietary interventions, also novel pharmacological interventions are in development that mimic the positive effects of the former.

Translation

Based on the important discoveries of biological mechanisms in the ageing process and possible intervention strategies, one crucial question is how to implement these new developments. What are the translational barriers? Which stakeholders need to be involved? How to communicate new treatments, diets, devices, and so on? How to convince end-users to use novel products? In order to explore these questions, we have used telehealthcare devices as a translational showcase.

Numerous telehealthcare devices exist from basic personal alarms and fall detectors through to complex monitoring devices for health-related vital signs. Such devices could be used to help individuals live at home for longer; however, they often go unused in practice.

Our study suggested that few telehealthcare devices were developed based on the principles of user-centred design. As a result, many were non-intuitive, with the majority of the focus group participants not recognising the purpose of the devices from their appearance alone. Expert interviewees argued that this lack of user-centred design resulted in unused products and products that failed to successfully meet end-user needs. Public and professional awareness campaigns are required with appropriate funding mechanisms for users to gain access to devices. The numerous barriers require systematic addressing, so device usage is promoted enabling individuals to live at home in good health for longer.

Implications for Policy and Practice

In the past decades, biogerontological research has contributed significantly to a better understanding of the ageing process that now provides the basis for the development of efficient intervention strategies suitable to increase healthy LE. On the one hand, these are established intervention strategies such as optimised nutrition and diet, regular physical activity, or prevention of infectious diseases through vaccination (Fig. 7.2). There is strong evidence for and consensus regarding the efficacy of these strategies. On the other hand, there are also novel intervention strategies in development, in particular novel pharmacological interventions (Fig. 7.2). These novel interventions aim to interfere with the ageing pathways that are nowadays better understood.

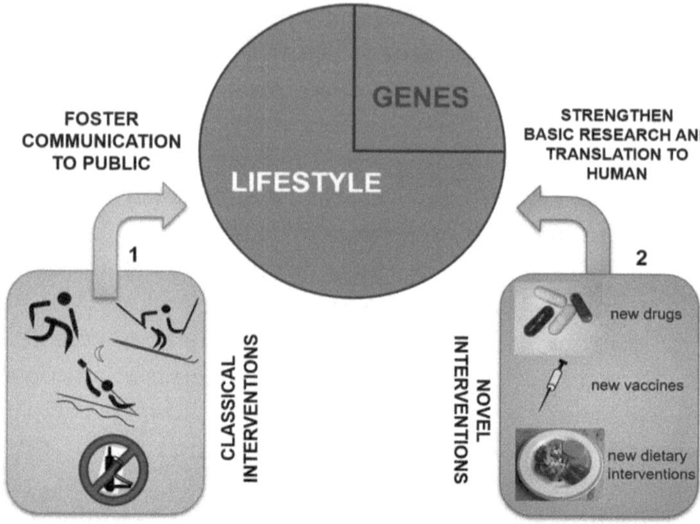

Fig. 7.2 Strategies to increase HLY. **(1) Established interventions** include nutrition, exercise, prevention of infectious diseases, no smoking/alcohol/drugs. **(2) Novel interventions** include dietary interventions, caloric restriction mimetics, clearance of senescent and damaged cells, mitohormetics, stem cells, drugs against inflammation, rejuvenation factors from blood, telomeres, and so on (capsules, healthy meal, pictograms: Institute for Biomedical Aging Research)

These two different branches of strategies require a different approach by policy for successful application in practice. For the established strategies with extensive evidence, broad and efficient communication to the public is needed firstly. Secondly, although the knowledge is transferred to the general publication, there is often a lack of adopting a healthy lifestyle by the informed subjects at the same time. Substantial research is still needed in this area to develop incentives but also regulations that would promote or even force the adoption of a healthy lifestyle. It is also specifically important to develop strategies to reach the socially disadvantaged already lagging behind in healthy LE with these established interventions.

Novel interventions still have to go a long developmental and regulatory path before their broad application in the population. However, these would then require less compliance and motivation for their successful application than the established strategies. In this area, politics is required to provide a suitable research and regulatory environment to facilitate the transition from preclinical data in animal models to clinical trials in humans and finally to interventions and treatments in the general population. Also, the scientific progress in this area needs to be efficiently disseminated to the relevant stakeholders and the general public. A key factor is the financial support that preclinical but also clinical research requires to speed up the progress in this field. And foremost a strategy is needed how to handle the long-term clinical trials that are likely required to prove the efficacy of novel interventions. Thus, one might need a scenario to accelerate the path to fulfil the regulatory requirements.

Further Research Priorities

As described in the previous sections, there is still much research needed in several areas.

- Biology and the biology of ageing in particular are complex. Although our understanding of the ageing process has significantly improved over the past decades, there is still much research needed to understand why there is such a big diversity in ageing, why some reach their 100th birthday healthy and others are threatened by disease and disability much earlier in life.
- Several intervention strategies have proven efficacy in animal models but lack clinical data in humans. Clinical research to prove the applicability of novel interventions in humans needs much more focus in the future.
- Established intervention strategies with proven evidence for their efficacy would be capable to increase HLY substantially at relatively low cost. However, novel communication strategies, incentives, and regulations need to be developed to promote and force the adoption of a healthy lifestyle in the general population.

Conclusion

LE is steadily increasing but HLY have difficulties to keep pace. Biogerontology can make a significant contribution to substantially extend the healthy period of our lives and thus booster active ageing.

- Accumulating evidence from experimental studies shows that ageing is not inevitably linked with the development of chronic diseases
- Only 20–25% of HLY is predetermined by genes; lifestyle and environment play a major role
- The age-associated accumulation of molecular and cellular damage can be prevented or greatly delayed by lifestyle interventions, for example, dietary manipulations
- Established strategies (e.g. nutrition, exercise, vaccination) require efficient communication to public and incentives for broad acceptance of a healthy lifestyle
- Novel strategies (e.g. dietary interventions, novel drugs, stem cells) need successful translation from the understanding of their molecular mechanisms to animal models to clinic
- To be successful, interventions need to be started early in life with a life-course perspective

References

Ackermann, R. T., Mulrow, C. D., Ramirez, G., Gardner, C. D., Morbidoni, L., & Lawrence, V. A. (2001). Garlic Shows Promise for Improving Some Cardiovascular Risk Factors. *Archives of Internal Medicine, 161*, 813–824.

Akazawa, N., Choi, Y., Miyaki, A., Tanabe, Y., Sugawara, J., Ajisaka, R., & Maeda, S. (2012). Curcumin Ingestion and Exercise Training Improve Vascular Endothelial Function in Postmenopausal Women. *Nutrition Research, 32*, 795–799.

Al-Azzawi, F., & Palacios, S. (2009). Hormonal Changes During Menopause. *Maturitas, 63*, 135–137.

Allan, G. M., Cranston, L., Lindblad, A., Mccormack, J., Kolber, M. R., Garrison, S., & Korownyk, C. (2016). Vitamin D: A Narrative Review Examining the Evidence for Ten Beliefs. *Journal of General Internal Medicine, 31*, 780–791.

Anderson, R. M., Shanmuganayagam, D., & Weindruch, R. (2009). Caloric Restriction and Aging: Studies in Mice and Monkeys. *Toxicologic Pathology, 37*, 47–51.

Anisimov, V. N., Berstein, L. M., Popovich, I. G., Zabezhinski, M. A., Egormin, P. A., Piskunova, T. S., Semenchenko, A. V., Tyndyk, M. L., Yurova, M. N., Kovalenko, I. G., & Poroshina, T. E. (2011a). If Started Early in Life, Metformin Treatment Increases Life Span and Postpones Tumors in Female SHR Mice. *Aging (Albany NY), 3*, 148–157.

Anisimov, V. N., Zabezhinski, M. A., Popovich, I. G., Piskunova, T. S., Semenchenko, A. V., Tyndyk, M. L., Yurova, M. N., Rosenfeld, S. V., & Blagosklonny, M. V. (2011b). Rapamycin Increases Lifespan and Inhibits Spontaneous Tumorigenesis in Inbred Female mice. *Cell Cycle, 10*, 4230–4236.

Baker, D. J., Wijshake, T., Tchkonia, T., LeBrasseur, N. K., Childs, B. G., van de Sluis, B., Kirkland, J. L., & van Deursen, J. M. (2011). Clearance of p16Ink4a-Positive Senescent Cells Delays Ageing-Associated Disorders. *Nature, 479*(7372), 232–236.

Bannister, C. A., Holden, S. E., Jenkins-Jones, S., Morgan, C. L., Halcox, J. P., Schernthaner, G., Mukherjee, J., & Currie, C. J. (2014). Can People with Type 2 Diabetes Live Longer Than Those Without? A Comparison of Mortality in People Initiated with Metformin or Sulphonylurea Monotherapy and Matched, Non-diabetic Controls. *Diabetes, Obesity and Metabolism, 16*, 1165–1173.

Baracco, G. J., Eisert, S., Saavedra, S., Hirsch, P., Marin, M., & Ortega-Sanchez, I. R. (2015). Clinical and Economic Impact of Various Strategies for Varicella Immunity Screening and Vaccination of Health Care Personnel. *American Journal of Infection Control, 43*, 1053–1060.

Barouki, R., Gluckman, P. D., Grandjean, P., Hanson, M., & Heindel, J. J. (2012). Developmental Origins of Non-communicable Disease: Implications for Research and Public Health. *Environmental Health, 11*, 1–9.

Bartke, A., Coschigano, K., Kopchick, J., Chandrashekar, V., Mattison, J., Kinney, B., & Hauck, S. (2001). Genes That Prolong Life: Relationships of Growth Hormone and Growth to Aging and Life Span. *The Journals of Gerontology. Series A, Biological Sciences and Medical Sciences, 56*, B340–B349.

Barzilai, N., Crandall, J. P., Kritchevsky, S. B., & Espeland, M. A. (2016). Metformin as a Tool to Target Aging. *Cell Metabolism, 23*, 1060–1065.

Baur, J. A., & Sinclair, D. A. (2006). Therapeutic Potential of Resveratrol: The in Vivo Evidence. *Nature Reviews. Drug Discovery, 5*, 493–506.

Baur, J. A., Pearson, K. J., Price, N. L., Jamieson, H. A., Lerin, C., Kalra, A., Prabhu, V. V., Allard, J. S., Lopez-Lluch, G., Lewis, K., Pistell, P. J., Poosala, S., Becker,

K. G., Boss, O., Gwinn, D., Wang, M., Ramaswamy, S., Fishbein, K. W., Spencer, R. G., Lakatta, E. G., Le Couteur, D., Shaw, R. J., Navas, P., Puigserver, P., Ingram, D. K., de Cabo, R., & Sinclair, D. A. (2006). Resveratrol Improves Health and Survival of Mice on a High-Calorie Diet. *Nature, 444*, 337–342.

Bengtsson, T., & Lindström, M. (2003). Airborne Infectious Diseases During Infancy and Mortality in Later Life in Southern Sweden, 1766–1894. *International Journal of Epidemiology, 32*, 286–294.

Bergmann, R. L., Diepgen, T. L., Kuss, O., Bergmann, K. E., Kujat, J., Dudenhausen, J. W., & Wahn, U. (2002). Breastfeeding Duration Is a Risk Factor for Atopic Eczema. *Clinical and Experimental Allergy, 32*, 205–209.

Berstein, L. M. (2012). Metformin in Obesity, Cancer and Aging: Addressing Controversies. *Aging (Albany NY), 4*, 320–329.

Bitto, A., & Kaeberlein, M. (2014). Rejuvenation: It's in Our Blood. *Cell Metabolism, 20*, 2–4.

Bitto, A., Ito, T. K., Pineda, V. V., Letexier, N. J., Huang, H. Z., Sutlief, E., Tung, H., Vizzini, N., Chen, B., Smith, K., Meza, D., Yajima, M., Beyer, R. P., Kerr, K. F., Davis, D. J., Gillespie, C. H., Snyder, J. M., Treuting, P. M., & Kaeberlein, M. (2016). Transient Rapamycin Treatment Can Increase Lifespan and Healthspan in Middle-Aged Mice. *Elife*, 5, pii: e16351.

Bjedov, I., & Partridge, L. (2011). A Longer and Healthier Life with TOR Down-Regulation: Genetics and Drugs. *Biochemical Society Transactions, 39*, 460–465.

Bjelakovic, G., Nikolova, D., Gluud, L. L., Simonetti, R. G., & Gluud, C. (2012). Antioxidant Supplements for Prevention of Mortality in Healthy Participants and Patients with Various Diseases. *The Cochrane Database of Systematic Reviews*, Cd007176.

Bjelakovic, G., Gluud, L. L., Nikolova, D., Whitfield, K., Wetterslev, J., Simonetti, R. G., Bjelakovic, M., & Gluud, C. (2014). Vitamin D Supplementation for Prevention of Mortality in Adults. *The Cochrane Database of Systematic Reviews*, Cd007470.

Blackburn, E. H., Epel, E. S., & Lin, J. (2015). Human Telomere Biology: A Contributory and Interactive Factor in Aging, Disease Risks, and Protection. *Science, 350*, 1193–1198.

Brondino, N., & Re, S. (2014). Curcumin as a Therapeutic Agent in Dementia: A Mini Systematic Review of Human Studies. *The Scientific World Journal, 2014*, 174282.

Brooks-Wilson, A. R. (2013). Genetics of Healthy Aging and Longevity. *Human Genetics, 132*, 1323–1338.

Campisi, J., & Robert, L. (2014). Cell Senescence, Role in Aging and Age-Related Diseases. *Interdisciplinary Topics in Gerontology, 39*, 45–61.

Caruso, C., Passarino, G., Puca, A., & Scapagnini, G. (2012). "Positive Biology": The Centenarian Lesson. *Immunity & Ageing, 9*, 5.

Carwile, J. L., & Michels, K. B. (2011). Urinary Bisphenol A and Obesity: NHANES 2003–2006. *Environmental Research, 111*, 825–830.

Colman, R. J., Beasley, T. M., Kemnitz, J. W., Johnson, S. C., Weindruch, R., & Anderson, R. M. (2014). Caloric Restriction Reduces Age-Related and All-Cause Mortality in Rhesus Monkeys. *Nature Communications, 5*, 3557.

Conboy, I. M., Conboy, M. J., & Rebo, J. (2015). Systemic Problems: A Perspective on Stem Cell Aging and Rejuvenation. *Aging (Albany NY), 7*, 754–765.

Conti, V., Izzo, V., Corbi, G., Russomanno, G., Manzo, V., De Lise, F., Di Donato, A., & Filippelli, A. (2016). Antioxidant Supplementation in the Treatment of Aging-Associated Diseases. *Frontiers in Pharmacology, 7*, 24.

Dai, D. F., Chiao, Y. A., Marcinek, D. J., Szeto, H. H., & Rabinovitch, P. S. (2014). Mitochondrial Oxidative Stress in Aging and Healthspan. *Longev Healthspan, 3*, 6.

Darnton-Hill, I., Nishida, C., & James, W. P. (2004). A Life Course Approach to Diet, Nutrition and the Prevention of Chronic Diseases. *Public Health Nutrition, 7*, 101–121.

de Cabo, R., Carmona-Gutierrez, D., Bernier, M., Hall, M. N., & Madeo, F. (2014). The Search for Antiaging Interventions: From Elixirs to Fasting Regimens. *Cell, 157*, 1515–1526.

De Haes, W., Frooninckx, L., Van Assche, R., Smolders, A., Depuydt, G., Billen, J., Braeckman, B. P., Schoofs, L., & Temmerman, L. (2014). Metformin Promotes Lifespan Through Mitohormesis Via the Peroxiredoxin Prdx-2. *Proceedings of the National Academy of Sciences, 111*, E2501–E2509.

Debost-Legrand, A., Warembourg, C., Massart, C., Chevrier, C., Bonvallot, N., Monfort, C., Rouget, F., Bonnet, F., & Cordier, S. (2016). Prenatal Exposure to Persistent Organic Pollutants and Organophosphate Pesticides, and Markers of Glucose Metabolism at Birth. *Environmental Research, 146*, 207–217.

Dewey, K. G., Heinig, M. J., & Nommsen-Rivers, L. A. (1995). Differences in Morbidity Between Breast-Fed and Formula-Fed Infants. *The Journal of Pediatrics, 126*, 696–702.

Dolan, G. P., Harris, R. C., Clarkson, M., Sokal, R., Morgan, G., Mukaigawara, M., Horiuchi, H., Hale, R., Stormont, L., Bechard-Evans, L., Chao, Y. S.,

Eremin, S., Martins, S., Tam, J. S., Penalver, J., Zanuzdana, A., & Nguyen-Van-Tam, J. S. (2012). Vaccination of Health Care Workers to Protect Patients at Increased Risk for Acute Respiratory Disease. *Emerging Infectious Diseases, 18*, 1225–1234.

Dolara, P., Bigagli, E., & Collins, A. (2012). Antioxidant Vitamins and Mineral Supplementation, Life Span Expansion and Cancer Incidence: A Critical Commentary. *European Journal of Nutrition, 51*, 769–781.

Doll, R., Peto, R., Boreham, J., & Sutherland, I. (2004). Mortality in Relation to Smoking: 50 Years' Observations on Male British Doctors. *BMJ, 328*, 1519.

Duijts, L., Jaddoe, V. W. V., Hofman, A., & Moll, H. A. (2010). Prolonged and Exclusive Breastfeeding Reduces the Risk of Infectious Diseases in Infancy. *Pediatrics, 126*, e18–e25.

Eisenberg, T., Knauer, H., Schauer, A., Buttner, S., Ruckenstuhl, C., Carmona-Gutierrez, D., Ring, J., Schroeder, S., Magnes, C., Antonacci, L., Fussi, H., Deszcz, L., Hartl, R., Schraml, E., Criollo, A., Megalou, E., Weiskopf, D., Laun, P., Heeren, G., Breitenbach, M., Grubeck-Loebenstein, B., Herker, E., Fahrenkrog, B., Frohlich, K.-U., Sinner, F., Tavernarakis, N., Minois, N., Kroemer, G., & Madeo, F. (2009). Induction of Autophagy by Spermidine Promotes Longevity. *Nature Cell Biology, 11*, 1305–1314.

Ernst, E. (2002). The Risk-Benefit Profile of Commonly Used Herbal Therapies: Ginkgo, St. John's Wort, Ginseng, Echinacea, Saw Palmetto, and Kava. *Annals of Internal Medicine, 136*, 42–53.

Feil, R., & Fraga, M. F. (2012). Epigenetics and the Environment: Emerging Patterns and Implications. *Nature Reviews. Genetics, 13*, 97–109.

Ferrucci, L., Cooper, R., Shardell, M., Simonsick, E. M., Schrack, J. A., & Kuh, D. (2016). Age-Related Change in Mobility: Perspectives From Life Course Epidemiology and Geroscience. *The Journals of Gerontology. Series A, Biological Sciences and Medical Sciences, 71*, 1184–1194.

Fiori, J. L., Shin, Y.-K., Kim, W., Krzysik-Walker, S. M., González-Mariscal, I., Carlson, O. D., Sanghvi, M., Moaddel, R., Farhang, K., Gadkaree, S. K., Doyle, M. E., Pearson, K. J., Mattison, J. A., de Cabo, R., & Egan, J. M. (2013). Resveratrol Prevents β-Cell Dedifferentiation in Nonhuman Primates Given a High-Fat/High-Sugar Diet. *Diabetes, 62*, 3500–3513.

Fisher, A., & Morley, J. E. (2002). Antiaging Medicine: The Good, the Bad, and the Ugly. *The Journals of Gerontology. Series A, Biological Sciences and Medical Sciences, 57*, M636–M639.

Fleg, J. L. (2012). Aerobic Exercise in the Elderly: A Key to Successful Aging. *Discovery Medicine, 13*, 223–228.

Fleischauer, A. T., & Arab, L. (2001). Garlic and Cancer: A Critical Review of the Epidemiologic Literature. *The Journal of Nutrition, 131*, 1032s–1040s.

Fontana, L. (2009). The Scientific Basis of Caloric Restriction Leading to Longer Life. *Current Opinion in Gastroenterology, 25*, 144–150.

Fontana, L., & Partridge, L. (2015). Promoting Health and Longevity Through Diet: From Model Organisms to Humans. *Cell, 161*, 106–118.

Fontana, L., Partridge, L., & Longo, V. D. (2010). Extending Healthy Life Span – From Yeast to Humans. *Science, 328*, 321–326.

Fontana, L., Adelaiye, R. M., Rastelli, A. L., Miles, K. M., Ciamporcero, E., Longo, V. D., Nguyen, H., Vessella, R., & Pili, R. (2013). Dietary Protein Restriction Inhibits Tumor Growth in Human Xenograft models. *Oncotarget, 4*, 2451–2461.

Forsyth, J. S. (1995). The Relationship Between Breast-Feeding and Infant Health and Development. *Proceedings of the Nutrition Society, 54*, 407–418.

Gavazzi, G. (2009). Influenza Vaccination for Healthcare Workers: From a Simple Concept to a Resistant Issue? *Aging Clinical and Experimental Research, 21*, 216–221.

Gergen, P. J., Mcquillan, G. M., Kiely, M., Ezzati-Rice, T. M., Sutter, R. W., & Virella, G. (1995). A Population-Based Serologic Survey of Immunity to Tetanus in the United States. *The New England Journal of Medicine, 332*, 761–766.

Giefing-Kroll, C., Berger, P., Lepperdinger, G., & Grubeck-Loebenstein, B. (2015). How Sex and Age Affect Immune Responses, Susceptibility to Infections, and Response to Vaccination. *Aging Cell, 14*, 309–321.

Gliemann, L., Schmidt, J. F., Olesen, J., Bienso, R. S., Peronard, S. L., Grandjean, S. U., Mortensen, S. P., Nyberg, M., Bangsbo, J., Pilegaard, H., & Hellsten, Y. (2013). Resveratrol Blunts the Positive Effects of Exercise Training on Cardiovascular Health in Aged Men. *The Journal of Physiology, 591*, 5047–5059.

Gluckman, P. D., Hanson, M. A., Cooper, C., & Thornburg, K. L. (2008). Effect of in Utero and Early-Life Conditions on Adult Health and Disease. *The New England Journal of Medicine, 359*, 61–73.

Gonzalez-Freire, M., de Cabo, R., Bernier, M., Sollott, S. J., Fabbri, E., Navas, P., & Ferrucci, L. (2015). Reconsidering the role of Mitochondria in Aging. *The Journals of Gerontology. Series A, Biological Sciences and Medical Sciences, 70*, 1334–1342.

Goodell, M. A., & Rando, T. A. (2015). Stem Cells and Healthy Aging. *Science, 350*, 1199–1204.

Groot, C., Hooghiemstra, A. M., Raijmakers, P. G., Van Berckel, B. N., Scheltens, P., Scherder, E. J., Van Der Flier, W. M., & Ossenkoppele, R. (2016). The Effect of Physical Activity on Cognitive Function in Patients with Dementia: A Meta-Analysis of Randomized Control Trials. *Ageing Research Reviews, 25*, 13–23.

Hale, M. W., Spencer, S. J., Conti, B., Jasoni, C. L., Kent, S., Radler, M. E., Reyes, T. M., & Sominsky, L. (2015). Diet, Behavior and Immunity Across the Lifespan. *Neuroscience and Biobehavioral Reviews, 58*, 46–62.

Hanson, M. A., & Gluckman, P. D. (2014). Early Developmental Conditioning of Later Health and Disease: Physiology or Pathophysiology? *Physiological Reviews, 94*, 1027–1076.

Harvie, M. N., Pegington, M., Mattson, M. P., Frystyk, J., Dillon, B., Evans, G., Cuzick, J., Jebb, S. A., Martin, B., Cutler, R. G., Son, T. G., Maudsley, S., Carlson, O. D., Egan, J. M., Flyvbjerg, A., & Howell, A. (2011). The Effects of Intermittent or Continuous Energy Restriction on Weight Loss and Metabolic Disease Risk Markers: A Randomized Trial in Young Overweight Women. *International Journal of Obesity, 35*, 714–727.

Heilbronn, L. K., Smith, S. R., Martin, C. K., Anton, S. D., & Ravussin, E. (2005). Alternate-Day Fasting in Nonobese Subjects: Effects on Body Weight, Body Composition, and Energy Metabolism. *The American Journal of Clinical Nutrition, 81*, 69–73.

Hercberg, S., Galan, P., Preziosi, P., Alfarez, M. J., & Vazquez, C. (1998). The Potential Role of Antioxidant Vitamins in Preventing Cardiovascular Diseases and Cancers. *Nutrition, 14*, 513–520.

Hypponen, E., Kenward, M. G., Virtanen, S. M., Piitulainen, A., Virta-Autio, P., Tuomilehto, J., Knip, M., & Akerblom, H. K. (1999). Infant Feeding, Early Weight Gain, and Risk of Type 1 Diabetes. Childhood Diabetes in Finland (DiMe) Study Group. *Diabetes Care, 22*, 1961–1965.

Jimenez-Gomez, Y., Mattison, J. A., Pearson, K. J., Martin-Montalvo, A., Palacios, H. H., Sossong, A. M., Ward, T. M., Younts, C. M., Lewis, K., Allard, J. S., Longo, D. L., Belman, J. P., Malagon, M. M., Navas, P., Sanghvi, M., Moaddel, R., Tilmont, E. M., Herbert, R. L., Morrell, C. H., Egan, J. M., Baur, J. A., Ferrucci, L., Bogan, J. S., Bernier, M., & de Cabo, R. (2013). Resveratrol Improves Adipose Insulin Signaling and Reduces the Inflammatory Response in Adipose Tissue of Rhesus Monkeys on High-Fat, High-Sugar Diet. *Cell Metabolism, 18*, 533–545.

Johannsson, G., Svensson, J., & Bengtsson, B. A. (2000). Growth Hormone and Ageing. *Growth Hormone & IGF Research, 10 Suppl B*, S25–S30.

Johnson, J. B., Summer, W., Cutler, R. G., Martin, B., Hyun, D. H., Dixit, V. D., Pearson, M., Nassar, M., Telljohann, R., Maudsley, S., Carlson, O., John, S., Laub, D. R., & Mattson, M. P. (2007). Alternate Day Calorie Restriction Improves Clinical Findings and Reduces Markers of Oxidative Stress and Inflammation in Overweight Adults with Moderate Asthma. *Free Radical Biology & Medicine, 42*, 665–674.

Johnson, S. C., Rabinovitch, P. S., & Kaeberlein, M. (2013). mTOR Is a Key Modulator of Ageing and Age-Related Disease. *Nature, 493*, 338–345.

Kaeberlein, M., Rabinovitch, P. S., & Martin, G. M. (2015). Healthy Aging: The Ultimate Preventative Medicine. *Science, 350*, 1191–1193.

Kaliman, P., Parrizas, M., Lalanza, J. F., Camins, A., Escorihuela, R. M., & Pallas, M. (2011). Neurophysiological and Epigenetic Effects of Physical Exercise on the Aging Process. *Ageing Research Reviews, 10*, 475–486.

Kramer, M. S., Matush, L., Vanilovich, I., Platt, R., Bogdanovich, N., Sevkovskaya, Z., Dzikovich, I., Shishko, G., & Mazer, B. (2007). Effect of Prolonged and Exclusive Breast Feeding on Risk of Allergy and Asthma: Cluster Randomised Trial. *BMJ, 335*, 815.

Kull, I., Bohme, M., Wahlgren, C. F., Nordvall, L., Pershagen, G., & Wickman, M. (2005). Breast-Feeding Reduces the Risk for Childhood Eczema. *The Journal of Allergy and Clinical Immunology, 116*, 657–661.

Laaksonen, D. E., Lindstrom, J., Lakka, T. A., Eriksson, J. G., Niskanen, L., Wikstrom, K., Aunola, S., Keinanen-Kiukaanniemi, S., Laakso, M., Valle, T. T., Ilanne-Parikka, P., Louheranta, A., Hamalainen, H., Rastas, M., Salminen, V., Cepaitis, Z., Hakumaki, M., Kaikkonen, H., Harkonen, P., Sundvall, J., Tuomilehto, J., & Uusitupa, M. (2005). Physical Activity in the Prevention of Type 2 Diabetes: The Finnish Diabetes Prevention Study. *Diabetes, 54*, 158–165.

Lafortune, L., Martin, S., Kelly, S., Kuhn, I., Remes, O., Cowan, A., & Brayne, C. (2016). Behavioural Risk Factors in Mid-Life Associated with Successful Ageing, Disability, Dementia and Frailty in Later Life: A Rapid Systematic Review. *PLoS One, 11*, e0144405.

Lanzieri, G. (2011). The Greying of the Baby Boomers. In Eurostat (Ed.), *Population and Social Conditions.* European Union: Eurostat – Statistics in Focus.

Lee, I. M., & Skerrett, P. J. (2001). Physical Activity and All-Cause Mortality: What Is the Dose-Response Relation? *Medicine and Science in Sports and Exercise, 33*, S459–S471; Discussion S493–S494.

Li, S., & Laher, I. (2015). Exercise Pills: At the Starting Line. *Trends in Pharmacological Sciences, 36*(12), 906–917.

Liu, K., Zhou, R., Wang, B., & Mi, M. T. (2014). Effect of Resveratrol on Glucose Control and Insulin Sensitivity: A Meta-Analysis of 11 Randomized Controlled Trials. *The American Journal of Clinical Nutrition, 99*, 1510–1519.

Longo, V. D., & Mattson, M. P. (2014). Fasting: Molecular Mechanisms and Clinical Applications. *Cell Metabolism, 19*, 181–192.

MacGillivray, D. M., & Kollmann, T. R. (2014). The Role of Environmental Factors in Modulating Immune Responses in Early Life. *Frontiers in Immunology, 5*, 434.

Mannick, J. B., Del Giudice, G., Lattanzi, M., Valiante, N. M., Praestgaard, J., Huang, B., Lonetto, M. A., Maecker, H. T., Kovarik, J., Carson, S., Glass, D. J., & Klickstein, L. B. (2014). mTOR Inhibition Improves Immune Function in the Elderly. *Science Translational Medicine, 6*, 268ra179.

Martin-Montalvo, A., Mercken, E. M., Mitchell, S. J., Palacios, H. H., Mote, P. L., Scheibye-Knudsen, M., Gomes, A. P., Ward, T. M., Minor, R. K., Blouin, M. J., Schwab, M., Pollak, M., Zhang, Y., Yu, Y., Becker, K. G., Bohr, V. A., Ingram, D. K., Sinclair, D. A., Wolf, N. S., Spindler, S. R., Bernier, M., & de Cabo, R. (2013). Metformin Improves Healthspan and Lifespan in Mice. *Nature Communications, 4*, 2192.

Mattison, J. A., Wang, M., Bernier, M., Zhang, J., Park, S. S., Maudsley, S., An, S. S., Santhanam, L., Martin, B., Faulkner, S., Morrell, C., Baur, J. A., Peshkin, L., Sosnowska, D., Csiszar, A., Herbert, R. L., Tilmont, E. M., Ungvari, Z., Pearson, K. J., Lakatta, E. G., & de Cabo, R. (2014). Resveratrol Prevents High Fat/Sucrose Diet-induced Central Arterial Wall Inflammation and Stiffening in Nonhuman Primates. *Cell Metabolism, 20*, 183–190.

Mattson, M. P., Allison, D. B., Fontana, L., Harvie, M., Longo, V. D., Malaisse, W. J., Mosley, M., Notterpek, L., Ravussin, E., Scheer, F. A., Seyfried, T. N., Varady, K. A., & Panda, S. (2014). Meal Frequency and Timing in Health and Disease. *Proceedings of the National Academy of Sciences of the United States of America, 111*, 16647–16653.

Mazumder, B., Almond, D., Park, K., Crimmins, E. M., & Finch, C. E. (2010). Lingering Prenatal Effects of the 1918 Influenza Pandemic on Cardiovascular Disease. *Journal of Developmental Origins of Health and Disease, 1*, 26–34.

McCarron, P., Smith, G. D., Okasha, M., & Mcewen, J. (2000). Blood Pressure in Young Adulthood and Mortality from Cardiovascular Disease. *Lancet, 355*, 1430–1431.

McPhee, J. S., French, D. P., Jackson, D., Nazroo, J., Pendleton, N., & Degens, H. (2016). Physical Activity in Older Age: Perspectives for Healthy Ageing and Frailty. *Biogerontology, 17*, 567–580.

Mercken, E. M., Carboneau, B. A., Krzysik-Walker, S. M., & de Cabo, R. (2012). Of Mice and Men: The Benefits of Caloric Restriction, Exercise, and Mimetics. *Ageing Research Reviews, 11*, 390–398.

Mercken, E. M., Crosby, S. D., Lamming, D. W., Jebailey, L., Krzysik-Walker, S., Villareal, D. T., Capri, M., Franceschi, C., Zhang, Y., Becker, K., Sabatini, D. M., de Cabo, R., & Fontana, L. (2013). Calorie Restriction in Humans Inhibits the PI3K/AKT Pathway and Induces a Younger Transcription Profile. *Aging Cell, 12*, 645–651.

Michel, J., Newton, J. L., & Kirkwood, T. L. (2008). Medical Challenges of Improving the Quality of a Longer Life. *JAMA, 299*, 688–690.

Miller, R. A., Harrison, D. E., Astle, C. M., Baur, J. A., Boyd, A. R., de Cabo, R., Fernandez, E., Flurkey, K., Javors, M. A., Nelson, J. F., Orihuela, C. J., Pletcher, S., Sharp, Z. D., Sinclair, D., Starnes, J. W., Wilkinson, J. E., Nadon, N. L., & Strong, R. (2011). Rapamycin, but Not Resveratrol or Simvastatin, Extends Life Span of Genetically Heterogeneous Mice. *The Journals of Gerontology. Series A, Biological Sciences and Medical Sciences, 66*, 191–201.

Monteverde, M., Noronha, K., & Palloni, A. (2009). Effect of Early Conditions on Disability Among the Elderly in Latin America and the Caribbean. *Population Studies (Camb), 63*, 21–35.

Morley, J. E., & Flaherty, J. H. (2002). It's Never Too Late: Health Promotion and Illness Prevention in Older Persons. *The Journals of Gerontology. Series A, Biological Sciences and Medical Sciences, 57*, M338–M342.

Mulvey, L., Sinclair, A., & Selman, C. (2014). Lifespan Modulation in Mice and the Confounding Effects of Genetic Background. *Journal of Genetics and Genomics, 41*, 497–503.

Nicklas, B. J., & Brinkley, T. E. (2009). Exercise Training as a Treatment for Chronic Inflammation in the Elderly. *Exercise and Sport Sciences Reviews, 37*, 165–170.

O'Toole, P. W., & Jeffery, I. B. (2015). Gut Microbiota and Aging. *Science, 350*, 1214–1215.

Oliveras-Lopez, M. J., Molina, J. J., Mir, M. V., Rey, E. F., Martin, F., & De La Serrana, H. L. (2013). Extra Virgin Olive Oil (EVOO) Consumption and Antioxidant Status in Healthy Institutionalized Elderly Humans. *Archives of Gerontology and Geriatrics, 57*, 234–242.

Paliy, O., Piyathilake, C. J., Kozyrskyj, A., Celep, G., Marotta, F., & Rastmanesh, R. (2014). Excess Body Weight During Pregnancy and Offspring Obesity: Potential Mechanisms. *Nutrition, 30*, 245–251.

Paricio Talayero, J. M., Lizán-García, M., Puime, Á. O., Muncharaz, M. J. B., Soto, B. B., Sánchez-Palomares, M., Serrano, L. S., & Rivera, L. L. (2006). Full Breastfeeding and Hospitalization as a Result of Infections in the First Year of Life. *Pediatrics, 118*, e92–e99.

Partridge, L. (2010). The New Biology of Ageing. *Philosophical Transactions of the Royal Society, B: Biological Sciences, 365*, 147–154.

Partridge, L. (2014). Intervening in Ageing to Prevent the Diseases of Ageing. *Trends in Endocrinology and Metabolism, 25*, 555–557.

Patel, K. R., Scott, E., Brown, V. A., Gescher, A. J., Steward, W. P., & Brown, K. (2011). Clinical Trials of Resveratrol. *Annals of the New York Academy of Sciences, 1215*, 161–169.

Pembrey, M., Saffery, R., Bygren, L. O., & Epidemiology, N. I. E. (2014). Human Transgenerational Responses to Early-Life Experience: Potential Impact on Development, Health and Biomedical Research. *Journal of Medical Genetics, 51*, 563–572.

Podmore, I. D., Griffiths, H. R., Herbert, K. E., Mistry, N., Mistry, P., & Lunec, J. (1998). Vitamin C Exhibits Pro-oxidant Properties. *Nature, 392*, 559.

Poland, G. A., Jacobson, R. M., & Ovsyannikova, I. G. (2009). Trends Affecting the Future of Vaccine Development and Delivery: The Role of Demographics, Regulatory Science, the Anti-vaccine Movement, and Vaccinomics. *Vaccine, 27*, 3240–3244.

Prato, R., Tafuri, S., Fortunato, F., & Martinelli, D. (2010). Vaccination in Healthcare Workers: An Italian Perspective. *Expert Review of Vaccines, 9*, 277–283.

Pucciarelli, S., Moreschini, B., Micozzi, D., De Fronzo, G. S., Carpi, F. M., Polzonetti, V., Vincenzetti, S., Mignini, F., & Napolioni, V. (2012). Spermidine and Spermine Are Enriched in Whole Blood of Nona/Centenarians. *Rejuvenation Research, 15*, 590–595.

Puterman, E., Lin, J., Krauss, J., Blackburn, E. H., & Epel, E. S. (2015). Determinants of Telomere Attrition Over 1 Year in Healthy Older Women: Stress and Health Behaviors Matter. *Molecular Psychiatry, 20*, 529–535.

Raffaghello, L., Lee, C., Safdie, F. M., Wei, M., Madia, F., Bianchi, G., & Longo, V. D. (2008). Starvation-Dependent Differential Stress Resistance Protects Normal but Not Cancer Cells Against High-Dose Chemotherapy.

Proceedings of the National Academy of Sciences of the United States of America, 105, 8215–8220.

Rattan, S. I. S. (2014). Aging Is Not a Disease: Implications for Intervention. *Aging and Disease, 5*, 196–202.

Robinson, S. M. (2015). Infant Nutrition and Lifelong Health: Current Perspectives and Future Challenges. *Journal of Developmental Origins of Health and Disease, 6*, 384–389.

Ruchat, S.-M., Bouchard, L., & Hivert, M.-F. (2014). Early Infant Nutrition and Metabolic Programming: What Are the Potential Molecular Mechanisms? *Current Nutrition Reports, 3*, 281–288.

Scudellari, M. (2015). Ageing Research: Blood to Blood. *Nature, 517*, 426–429.

Seals, D. R., & Melov, S. (2014). Translational Geroscience: Emphasizing Function to Achieve Optimal Longevity. *Aging (Albany NY), 6*, 718–730.

Sohal, R. S., & Forster, M. J. (2014). Caloric Restriction and the Aging Process: A Critique. *Free Radical Biology & Medicine, 73*, 366–382.

Spindler, S. R. (2005). Rapid and Reversible Induction of the Longevity, Anticancer and Genomic Effects of Caloric restriction. *Mechanisms of Ageing and Development, 126*, 960–966.

Steves, C. J., Spector, T. D., & Jackson, S. H. (2012). Ageing, Genes, Environment and Epigenetics: What Twin Studies Tell Us Now, and in the Future. *Age and Ageing, 41*, 581–586.

Tchkonia, T., Zhu, Y., Van Deursen, J., Campisi, J., & Kirkland, J. L. (2013). Cellular Senescence and the Senescent Secretory Phenotype: Therapeutic Opportunities. *The Journal of Clinical Investigation, 123*, 966–972.

Todd, C., & Skelton, D. (2004). What Are the Main Risk Factors for Falls Amongst Older People and What Are the Most Effective Interventions to Prevent These Falls? *Health Evidence Network report* [Online]. Available: http://www.euro.who.int/document/E82552.pdf. Accessed 5 Apr 2004.

Toschke, A. M., Vignerova, J., Lhotska, L., Osancova, K., Koletzko, B., & von Kries, R. (2002). Overweight and Obesity in 6- to 14-Year-Old Czech Children in 1991: Protective Effect of Breast-Feeding. *The Journal of Pediatrics, 141*, 764–769.

Tremblay, M. S., Gray, C., Babcock, S., Barnes, J., Bradstreet, C. C., Carr, D., Chabot, G., Choquette, L., Chorney, D., Collyer, C., Herrington, S., Janson, K., Janssen, I., Larouche, R., Pickett, W., Power, M., Sandseter, E. B., Simon, B., & Brussoni, M. (2015). Position Statement on Active Outdoor Play. *International Journal of Environmental Research and Public Health, 12*, 6475–6505.

Vafeiadi, M., Roumeliotaki, T., Myridakis, A., Chalkiadaki, G., Fthenou, E., Dermitzaki, E., Karachaliou, M., Sarri, K., Vassilaki, M., Stephanou, E. G., Kogevinas, M., & Chatzi, L. (2016). Association of Early Life Exposure to Bisphenol A with Obesity and Cardiometabolic Traits in Childhood. *Environmental Research, 146*, 379–387.

Vaiserman, A. (2011). Early-Life Origin of Adult Disease: Evidence from Natural Experiments. *Experimental Gerontology, 46*, 189–192.

Varady, K. A., Bhutani, S., Church, E. C., & Klempel, M. C. (2009). Short-Term Modified Alternate-Day Fasting: A Novel Dietary Strategy for Weight Loss and Cardioprotection in Obese Adults. *The American Journal of Clinical Nutrition, 90*, 1138–1143.

Vaupel, J. W., Carey, J. R., & Christensen, K. (2003). It's Never Too Late. *Science (New York, N.Y.), 301*, 1679–1681.

Verdin, E. (2015). NAD+ in Aging, Metabolism, and Neurodegeneration. *Science, 350*, 1208–1213.

Viner, R. M., Coffey, C., Mathers, C., Bloem, P., Costello, A., Santelli, J., & Patton, G. C. (2011). 50-Year Mortality Trends in Children and Young People: A Study of 50 Low-Income, Middle-Income, and High-Income Countries. *The Lancet, 377*, 1162–1174.

von Kries, R., Koletzko, B., Sauerwald, T., Von Mutius, E., Barnert, D., Grunert, V., & Von Voss, H. (1999). Breast Feeding and Obesity: Cross Sectional Study. *BMJ, 319*, 147–150.

Von Werder, K. (1999). The Somatopause Is No Indication for Growth Hormone Therapy. *Journal of Endocrinological Investigation, 22*, 137–141.

Wakeford, R., & Bithell, J. F. (2015). Childhood Cancer – The Role of Birthweight and Antenatal Radiography. *International Journal of Epidemiology, 44*, 1741–1743.

Walford, R. L., Mock, D., Verdery, R., & Maccallum, T. (2002). Calorie Restriction in Biosphere 2: Alterations in Physiologic, Hematologic, Hormonal, and Biochemical Parameters in Humans Restricted for a 2-Year Period. *The Journals of Gerontology. Series A, Biological Sciences and Medical Sciences, 57*, B211–B224.

Wang, Y., & Hekimi, S. (2015). Mitochondrial Dysfunction and Longevity in Animals: Untangling the Knot. *Science, 350*, 1204–1207.

Wang, X. S., Armstrong, M. E., Cairns, B. J., Key, T. J., & Travis, R. C. (2011). Shift Work and Chronic Disease: The Epidemiological Evidence. *Occupational Medicine (London), 61*, 78–89.

Wang, I. F., Guo, B. S., Liu, Y. C., Wu, C. C., Yang, C. H., Tsai, K. J., & Shen, C. K. (2012). Autophagy Activators Rescue and Alleviate Pathogenesis of a

Mouse Model with Proteinopathies of the TAR DNA-Binding Protein 43. *Proceedings of the National Academy of Sciences of the United States of America, 109*, 15024–15029.

Who Regional Office For Europe. (2004). *Physical Activity and Health in Europe: Evidence for Action* [Online]. Copenhagen: WHO Regional Office for Europe. Available: http://www.euro.who.int/de/what-we-publish/abstracts/physicalactivity-and-health-in-europe-evidence-for-action

Wilkinson, J. E., Burmeister, L., Brooks, S. V., Chan, C. C., Friedline, S., Harrison, D. E., Hejtmancik, J. F., Nadon, N., Strong, R., Wood, L. K., Woodward, M. A., & Miller, R. A. (2012). Rapamycin Slows Aging in Mice. *Aging Cell, 11*, 675–682.

Willcox, B. J., & Willcox, D. C. (2014). Caloric Restriction, Caloric Restriction Mimetics, and Healthy Aging in Okinawa: Controversies and Clinical Implications. *Current Opinion in Clinical Nutrition and Metabolic Care, 17*, 51–58.

Yi, Z., Gu, D., & Land, K. C. (2007). The Association of Childhood Socioeconomic Conditions with Healthy Longevity at the Oldest-Old Ages in China. *Demography, 44*, 497–518.

Yuan, C., Gaskins, A. J., Blaine, A. I., et al. (2016). Association Between Cesarean Birth and Risk of Obesity in Offspring in Childhood, Adolescence, and Early Adulthood. *JAMA Pediatrics*, e162385.

Zouboulis, C. C., & Makrantonaki, E. (2012). Hormonal Therapy of Intrinsic Aging. *Rejuvenation Research, 15*, 302–312.

8

Technology for All

Sebastian Merkel, Rolf G. Heinze, Josef Hilbert, and Gerhard Naegele

Introduction

The built and technical environments cover a huge variety of different aspects, ranging from assistive technologies to the housing situation and public transport. The environment which people live in accounts for

S. Merkel (✉) • J. Hilbert
Institute for Work and Technology, Westphalian University of Applied Sciences, Gelsenkirchen, Germany
e-mail: merkel@iat.eu; hilbert@iat.eu; http://www.iat.eu; http://www.iat.eu

R. G. Heinze
Ruhr-University Bochum, Bochum, Germany
e-mail: rolf.heinze@rub.de; https://www.sowi.rub.de/personal/heinze.html.en

G. Naegele
Research Association for Gerontology, TU Dortmund University, Dortmund, Germany
e-mail: ffg@institut-fuer-gerontologie.de; https://www.tu-dortmund.de/uni/en/Home/

A. Walker (ed.), *The Future of Ageing in Europe*,
https://doi.org/10.1007/978-981-13-1417-9_8

variation in the way they age—not only according to physical health, but also to personal satisfaction and the overall quality of life. This implies that the process of ageing can be positively supported by (age friendly) environments. MoPAct WP7 (built and technological environment) is concentrated on analysing the developments, trends, and challenges in the following three different areas: information and communication technologies (ICT), housing, and mobility.

ICT, housing, and mobility all play a major role within the context of ageing, each area in its own as well as directly linked to one another. The research was conducted within five European countries (Finland, Germany, Poland, Spain, and the United Kingdom). Our research aims can be summarized as follows:

- Identifying and analysing innovative ICT, mobility, and housing trends which are addressed to meet the needs of elderly in each of the countries involved;
- Case studies of best practices and their multi-level review and validation;
- Policy implications and recommendations.

In the literature, a variety of definitions of ICT can be found, leading to a broad understanding of this field. We mainly focused on *assistive technologies*, such as telecare products and, with respect to housing, active/ambient-assisted living. While 'basic' solutions, like social alarm systems, have widely been adapted in multiple countries, the vast majority of eHealth solutions have not gone beyond the state of pilot projects. In the five countries, some differences were found: comparing the overall availability of ICT, Spain and Poland fall behind the rest—especially compared to Finland and the United Kingdom. The same is true for the implementation of telecare. However, though there are differences between those countries, all show a common lack of considering the diffusion of assistive technologies.

In the field of *housing*, we identified four fields of action: active/ambient assisted living, shared housing, accessibility, and age-friendly neighbourhoods. The relevant stakeholders, be it politicians, local administrations, welfare associations, private landlords, or the housing industry and civil society organizations, will have to address these four fields of action in order to improve the housing situation of older people in a way that fosters active ageing. We found great variation between and inside the observed countries regarding the needs and resources of older people. Dimensions of this variance include the state of the dwelling (e.g. regarding accessibility), the local infrastructure (e.g. access to medical care), the ownership status, and the financial resources of the owner and/or tenant. Possible answers to the resulting challenges have to take into account these differences as well as the institutional and legal framework in the respective countries and regions.

Mobility is very important for (older) people in maintaining their independency and involvement in society. It can be seen as one of the fundamental preconditions to practice active ageing in later life. This is particularly true in both rural areas and for those groups of older people who are partially or totally dependent on public transport options, for example, because private transport options are not available (no own cars, financial need or restricted physical, and/or psychical constitution). Motives for older people to be mobile are to deal with everyday life, care for health, spend leisure time, as well as have a social life with family, friends, and neighbours and being socially integrated. How people use public or private transport varies greatly depending on how healthy they are, on how much they can afford to spend for transportation, and what attitude older people have towards public transport. Thus, key determinants for mobility in older age are cognitive, psychosocial, physical, environmental, and financial influences. Due to an overall improved health, more travelling options and better foreign language skills, older people today are more active than former generations. To ensure the continuous social inclusion of people as they age and to support an independent life at home as long as possible, the growing heterogeneity of older people along with their mobility and public transport needs must be taken into account. Affordability, availability, accessibility, and acceptability are the most basic issues to accomplish this, followed by security, information, reliableness, and cleanliness.

The Key Challenge: ICTs, Age-Friendly Housing, and Mobility for All?

ICT

While modern ICTs, such as smartphones or tablets, are often brought into connection with younger generations of users, this view has started to change within the last years. ICTs play a continuously important role for all ages, including older persons: for communicating with friends and family (e.g. via videoconferencing), in health and social care (e.g. by sending data directly to the doctor), at home (e.g. by installing smart home applications that automatically regulate the heating), with respect to mobility (e.g. driver assistance systems in cars) or simply during the leisure time (e.g. playing games online). Against the background of ageing societies and its consequences, modern ICT-based technologies have the potential to help persons living independently at home, support (older) persons with everyday tasks, as well as lowering costs of health and social care services and fostering economic growth.

The interplay of technology and ageing—ICTs are only one part—has been studied for almost 30 years. Bringing together several academic and practical disciplines, the field of 'gerontechnology' emerged in the late 1980s (van Bronswijk et al. 2009). The aim of the discipline is to provide 'an optimal technical environment for all ageing and old people up to a high age' (Bouma 2009).

Many of these hopes and aims, however, could not (yet) be realized due to severe barriers in different areas of implementation and many promising products and services have not been able to successfully compete on the market. In particular, 'high-tech' assistive devices, like ICT-based products and services, lack behind expectations of developers and policymakers.

Technical solutions will only succeed and be accepted by the market if a significant 'added value' for the end-user is created, which primarily lies in raising the quality of life of different subgroups in different life situations and environmental settings. Furthermore, technical solutions must be coordinated and linked to the offers and services of the traditional providers of ICT, health care industry, housing industry, public and private mobility provider, as well as social services. 'Strategic alliances' have to be incentivized and created mainly on the local level where the need of older people is directly addressed and where immediate solutions are looked for. The supply of technical assistance systems for the elderly is different in the countries studied. Established welfare states are more developed in terms of technical assistance in health care, housing and/or mobility, although there are differences between these countries. While only individual projects offer interesting prospects in Germany, over 10 per cent of seniors in the Netherlands or Scandinavian countries are already part of such support systems.

Comparable country examples show the necessity for the implementation of integrated supplies and solutions between the health care and the housing sector, which allows to a certain degree to see the private household as the third pillar of the health and care provision systems. For example, more than 2000 homes with sensors were provided in the Scottish county of West Lothian between 2002 and 2006. A simple social alarm system that transmits the data from portable crash sensors, motion, smoke, and gas detectors in an emergency to a call centre was installed. The technology was able to reduce the average length of stay in nursing homes of the district from 38 to less than ten months. The technical support costs less than half as much as the accommodation in a nursing home. After the end of the project, the Scottish government promoted the expansion of the age technology (Gast 2013; and for the international spread of telemedicine procedures among others van den Berg et al. 2015).

By networking personalized data, the domestic living environment appears to be a very interesting application for Internet solutions. In particular, emergency call systems, sensor mats for fall hazards, online office hours as well as applications in the field of tele-health monitoring (as in chronic diseases such as diabetes or heart failure) are of great importance to the housing for older people. Just when preserving the independence of ageing people is the focus, new solution formats can be offered by (mobile) Internet applications. Sensor-based ICT systems deliver custom service functions that can fulfil the specific needs (as both facing emergencies also for control of household appliances and to maintain social contacts through Internet platforms).

Health apps and 'wearables' (Fitness bracelets, watches, etc.) show how two basic socio-economic and cultural trends—individualization and digitization—meet and mutually reinforcing. An example can be demonstrated with reference to the 'quantified self' movement, which so far is practised primarily in the area of health, including the measurement of vital signs and physical activities, such as number of steps per day and so on by body worn digitized and miniaturized measuring devices that transmit their data to the storage systems over wireless networks (Lupton 2016; Heinze 2016 and the contributions in Selke 2016). Thus, the sensors can also be used to increase the quality of life of older people. Intelligent sensor technologies coupled with telemedical applications allow elderly people to fulfil their desire, as long as possible, to live independently. All of these examples show the smooth transition of technical assistance systems for telemedicine applications.

This topic is gaining in importance right now due to digitalization. About one-third of all adult Europeans are using the web for searching information on health. Independent, qualified patient survey portals will grow even more in the future. Already, the number of websites dealing with health and medicine issues is estimated to involve more than 100,000, with several hundred German-speaking medical and health portals. What is needed though is a clear evaluation system of such portals in order to improve the orientation for the user.

The increasing use of technologies to promote independent living in old age is boosted by the fact that the number of healthy years in old age has grown considerably. The state of health in old age has improved and the life expectancy in Western and Central Europe has increased rapidly since the late nineteenth century, although among European Union (EU) member states a clear East-West division is still persisting. Among Central and Eastern European (CEE) countries, this is particularly true for Poland, which SHARE data report the highest risk of diseases among older adults in the countries incorporated (this is also reported for Czechia). However, the relatively bad health status of older workers in CEE countries should not be (only) misinterpreted as result of political neglect. Health is affected by long-term factors, and CEE's current older people spent a significant share of their lives in pre-1990 times with worse conditions for health in paid work, as exemplarily emphasized in the WP3 Polish country report (Naegele and Bauknecht 2017) referred to in Chap. 5.

Various scientific studies show technology alone should not be the key centre of attention. The deficits in the professional care system should be the starting point for intervention. Therefore, the elderly affected need social assistance first and must then be integrated into a local care system with technological assistances. To link technologies with human and social assistance is one of the prime examples for socially innovative solutions in ageing societies (Heinze and Naegele 2013). However, the support of health and care, independent living and mobility for all, particularly in the local environment and at home, needs cooperation across the borders of economic sectors and institutions (from the real estate industry to health insurance).

There is plenty of empirical evidence available that ICT-based technology and infrastructure can help to support (elderly) people in their daily lives (Heinze et al. 2011; Heinze 2015, 2016). Despite this potential, the uptake of promising products and services shows significant differences among countries and groups of end-users, revealing that multiple variables and promising innovations have not been able to unfold their potentials (yet). Theoretically, there might be 'light at the end of the tunnel' if one considers significantly increased number of 'silver surfers' in recent years (Oswald and Wahl 2016). However, these numbers differ greatly among countries. Taking Internet use as an example, Fig. 8.1 compares the daily Internet use according to age in five European countries. It shows that the Internet use has increased continuously within the last years. While in the United Kingdom or Finland already a large amount of persons aged 65+ uses the Internet on a regular basis, this is not true for Poland or Spain.

Our research confirmed existing knowledge about socially biased differences in end-user-acceptance and take-up-rates: even though more and more persons in old age go online, this is not true considering the use of other (ICT-based) technology—in particular assistive devices like telehealth or telecare lack of use (Merkel and Enste 2015). EU

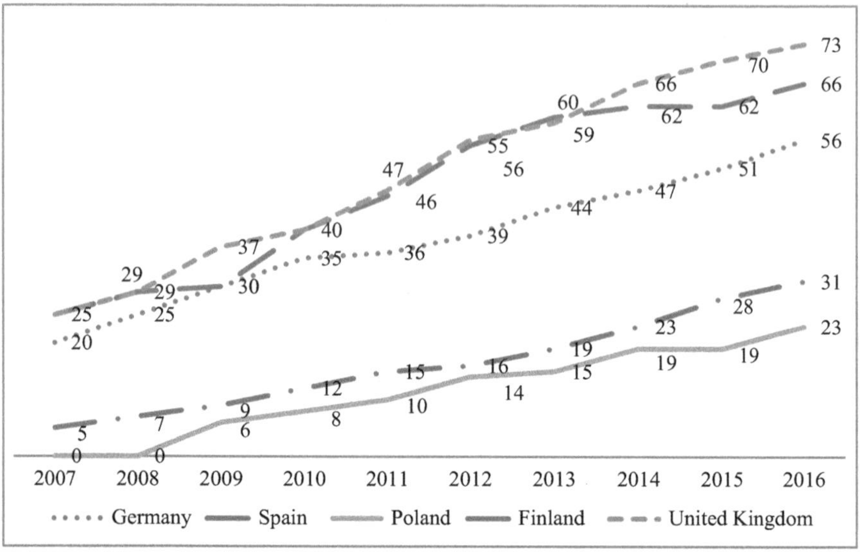

Fig. 8.1 Frequency of Internet access by individuals aged 65–74 years old: once a week (including every day). (Source: Eurostat [isoc_ci_ifp_fu])

wide, acceptance rates mirror a deepening well-known *social division* in old age (Ehlers and Naegele 2017), with particularly negative consequences for the very old and/or older persons with lower socio-economic status, both identified through research as especially in need. Moreover, next to the well-known factors of social segregation, the immediate benefits can be seen as a decisive factor to raise take up rates, although very often the 'added value' of conducive ICT is not primarily recognized by the elderly persons in need concerned but is dependent on the understanding of their immediate private and professional social environment (e.g. relatives, neighbours, professional helpers). Therefore, they have to be seen as second 'target group' of policy recommendations in terms of dissemination 'good practices'. On the other hand, experts assume that the striking clearly socially selective resistances will largely degrade gradually in the coming years, although further education measures and low-threshold offers of use are of great importance. The future elderly will differ significantly from today in terms of openness (Heinze et al. 2011).

Housing

Although *housing* gains in importance for older people, it is only one part of their life situation. In general, many of them live an active life in which not only family and hobbies but also voluntary services play a major role. The observable decrease in social networks in old age does not imply a decreasing satisfaction with life in general. On average, people aged 65–85 years spend five days a week with activities outside the private home, one third (32 per cent) even daily. Almost one out of two 75- to 79-year-olds drives a car (1985: one out of ten).

Housing offers that comprise a smooth transition from complete independence to graded caring and nursing assistance are still at the beginning stages. Such projects are often initiated by persons affected and are mainly implemented in new buildings. Often only a small group of people can afford them because of their high costs. The technological systems and home-care solutions employed in these cases are, however, not always adapted to the needs of the inhabitants. In future, the maxim should always be that technology serves the inhabitants (and not the other way around) and that it should be necessary and affordable. That said, ICTs open up completely new opportunities for assisting patients and clients in their own living environments, for example, by means of the remote monitoring of vital parameters, medical (and other) expertise, and—if necessary—by ensuring rapidly initiated remedial measures and interventions. Furthermore, they can help to make households barrier-free and adaptable so that several handicaps and limitations can be compensated.

The German long-term care (LTC) system offers provisions for technically adapting apartments and houses in case of being needy of care; however, experts criticize the restricted access to the LTC provisions (the non-officially recognized LTC-need is excluded), too small financial provisions, missing business models and quality criteria for the local building trade (Heinze et al. 2011). Optimally designed housing can also—in terms of a preventive technology and service deployment—contribute to putting off the need for help. Moreover, the Internet offers clients and patients abundant opportunities to inform themselves and to profit from an exchange with other interested and concerned persons. Although new technological options for active ageing are of great importance, the structural requirements for independent living for older people must also be considered. These are mostly not available in the Eastern European countries in particular, and even in highly developed welfare states such as Germany there are problems which mainly refer to rural areas as well to socially disadvantaged parts of urban areas. While many apartments have recently been modernized, there are still huge challenges in the field of housing in old age.

Such deficits could easily be improved by small changes in construction and lead to an improvement of life quality for people with health problems (e.g. deconstruction of barriers). In addition, it would be possible to make independent living in one's own apartment or house possible. This can be explained by the fact that the physical and cognitive skills change with age and environmental factors for the subjective well-being gain in importance. Also, with regard to independent living in old age avoiding falls is the main factor. Aside from health limitations, risk factors are those of the spatial environment such as tripping hazards in the apartment, lack of space for movement in the bathroom, and bad lighting. Many environmental factors such as these can easily be improved by simple building measures. Safety in old age is becoming a major issue and offers plenty of links for additional measures and service concepts. However, the current debate on housing in old age does not only focus on the design of the living space. Further spheres of activity are:

- Development of new cooperative dwelling forms (from assisted living to elderly flatshare).
- Setting up of an infrastructure in the local housing environment fitting the needs of older people including the creation of public meeting places.
- Promotion of voluntary services.
- Development of assistive technologies.

Mobility

Mobility is one of the key factors helping people to maintain their independence and involvement in society. However, the understanding of mobility varies contextually and can cover topics like physical mobility, driving, migration, occupational or holiday mobility. Our research focused mainly on mobility in terms of public transport. Mobility can be regarded as a basic need in (modern) societies, but for elderly people mobility can be quite a challenge and each step of the mobility chain could be a potential obstacle. Mobility in older age involves several dimensions. Key determinants are cognitive, psychosocial, physical, environmental, and financial influences (Webber et al. 2010); cognitive and/or physical impairments, such as declined hearing and vision or limitations in the musculoskeletal system increase with age and might hinder mobility. However, due to improved health, more travelling options, and better foreign language skills, older people today are more active than former generations at comparable ages. Another strong influence on mobility is the number of people living in one household, as those living alone tend to be less mobile than people living with a partner. In addition, financial aspects directly influence mobility and thus social inclusion since low economic resources limit the options for activities (Fiedler 2007; Follmer et al. 2010). To ensure the continuous social inclusion of people as they age and to support an independent life at home as long as possible, mobility and public transport needs must take into account also barrier-free mobility options.

The transport needs of older adults can be summarized under the four 'As' of public transport (DfT 2012a): availability (e.g. in rural areas), acceptability (e.g. for older people, being accustomed to drive an own car), accessibility (mainly barrier-freedom), and affordability (e.g. too high prices with excluding effects). We analysed mobility and transport policies and practices in Finland, Germany, Poland, Spain, and the United Kingdom, all with different states of development. The United Kingdom and Finland were placed at the top positions of 'good practices', whereas Poland and Spain were more programmatically on the way. Germany was ranked somewhere in-between.

In *Finland,* the transport policy is part of the economic, employment, and innovation policy. Being a country with long distances and dependent on exports, a good transport system is essential. Bus transport is the main mode of public transport. Due to high costs, accessibility cannot be provided in all travel centres of Finland and thus some travel needs must be fulfilled by special arrangements. Instead of providing special solutions for people with impairments, the primary goal is to promote a safe traffic environment and high-quality services which are suitable for everyone (MTC 2008).

In *Germany*, transport issues are and have been highly relevant for the Federal German Government for many decades; however, the reality partly shows the opposite: The recently published seventh Federal Report on the situation of older people in Germany identifies the spheres of 'mobility' and 'barrier freedom' as (among four others) those with the strongest need for action particularly at the local level (BMFSFJ 2016). For example, only 50 per cent of German train stations are barrier free. Currently, a new focus on the implications of demographic change on traffic, including features such as presumably more people being dependent on supportive services, more one-person-households and more individual motorized traffic emerging have priority on the policy agenda. Further factors of influence are the growing heterogeneity of lifestyles and regional differences. The current strategy for innovation in Germany ('Hightech-Strategie 2020') follows an integrative approach and contains so called 'future projects' (e.g. self-determined living in old age). Within this strategy mobility is seen as both one out of five global challenges as well as a key precondition for social inclusion. The focus is on technological solutions, barrier-free mobility offers, and assistant technologies for motoring and travel information. Local public transport is seen as an alternative to driving an own car; technical systems and intelligent services that adapt to the wishes and capabilities of the older generation in manifold life situations should be developed. Mobility and transport in *Poland* are mainly discussed on a programmatic level as steps to be taken on the way to reach the civilization development level and standard of living in Western European countries. An efficient transport system is seen as one facilitator of economic growth. The most significant objective is to improve the transport accessibility and transport quality in order to better conditions of living and to remove barriers for economic development. More specifically and directly targeting older persons, a priority is seen in improving the quality of transport in cities. In all, public transport is seen as an alternative to individual automobile traffic being reached through improved competitiveness of public transport, compared to individual transport and improved conditions for pedestrian and bicycle traffic with special emphasis on meeting the needs of disabled users (Ministry of Infrastructure 2005).

Also, in *Spain* a programmatic approach is dominating the public discussion on mobility and traffic; 'intelligent, sustainable and integrated transport' is said to be required. Priorities are set for technologies for reducing CO^2 emissions, developing intelligent cities, improving traffic flow, and enhancing accessibility for disabled persons (MEC 2013). Both equitable conditions of accessibility throughout the country and the importance of guaranteed universal minimum access to public services such as healthcare and social assistance for all citizens are emphasized. This includes adaptations of architectural barriers, walkways, and accessibility to public transportation for social groups with disabilities and priorities. Recently published policy papers give first priority to the protection of vulnerable groups, mainly those with reduced mobility and without access to a car: children, women, people with disabilities or low income, older people, and those who simply do not want to be dependent on cars. However, in practice, only few Spanish cities have a medium- or long-term urban mobility plan.

Innovative public transport according to the needs of older people with reduced mobility runs like a golden thread through all policy papers in the context of mobility and transport in the United Kingdom. The United Kingdom started comparatively early (in 2000) focussing on older people and mobility in rural areas, emphasizing the risk of social isolation and also aiming at supporting older people to stay at home and have a good health service. The difficulty to access services can be a particular problem for older people and is foremost seen as a transport issue (DETR 2000). Older people are officially identified as a vulnerable group whose needs have to be sufficiently drawn out to avoid exclusion, particularly through rural transport solutions and local public and community-based transport (DEFRA 2004). The policy 'Making transport more accessible to all' of 2012 (https://www.gov.uk/government/policies/making-transport-more-accessible-to-all) is an example the mainstreaming process and focuses primarily on access to buses, coaches, trains, and taxis for disabled people. One measure explicitly focusing on older people can be found in guidance and funding for community and voluntary transport (DfT 2012a). A few years ago, the charity age United Kingdom identified several transport barriers of concern for older people (e.g. physically inaccessible transport vehicles, the pedestrian environment, safety concerns, and attitudes of transport staff). These barriers are addressed in the 'Transport Solutions for Older People'. Examples are door to door transport (to improve availability) or travel training and mentoring (to improve acceptability) (DfT 2012b). Furthermore in 2012, the Department for Transport published the policy statement 'Green Light for Better Buses' that promotes concessionary travel in England for older and disabled people to give them greater freedom for visiting family and friends as well as reaching facilities within and outside their local area (DfT 2012c).

In the 'European Quality of Life Survey' (EQLS), which is carried out every four years in all 27 member states on behalf of Eurofound, 1000–3000 persons per country are interviewed face-to-face about employment, income, education, housing, family, health, work–life balance, life satisfaction, and perceived quality of society. The reasons for difficulties of access to doctors are explored and the distance to healthcare is another subcategory. The image of public services, including public transport, is an intensely questioned area and member states are compared. Age groups are not explicitly depicted but younger (18–24 years) are contrasted with older (50–64 years) (Eurofound 2012). In the following two figures based on the data of the EQLS, the access to and the quality of public transport for older people is analysed for the target countries Finland, Germany, Spain, Poland, and the United Kingdom. The access is described as being most easy in the United Kingdom, followed by Spain. It is rated as quite easy for the majority of regions in Germany but is still classified as only moderate in Poland. It stands out that Finish interviewees were evaluating the access to public transport facilities, as it seems that difficult access to public transport in Finland leads to disuse (Figs. 8.2 and 8.3). The quality of public transport is again rated highest by the respondents from the United Kingdom, followed by Spain. Finland's and Germany's quality of public transport is also mostly rated as high. Polish people evaluated the transport quality to a larger extent than the other countries as poor or very poor. This is in line with the Polish national policy priority to catch up with the Western European development level.

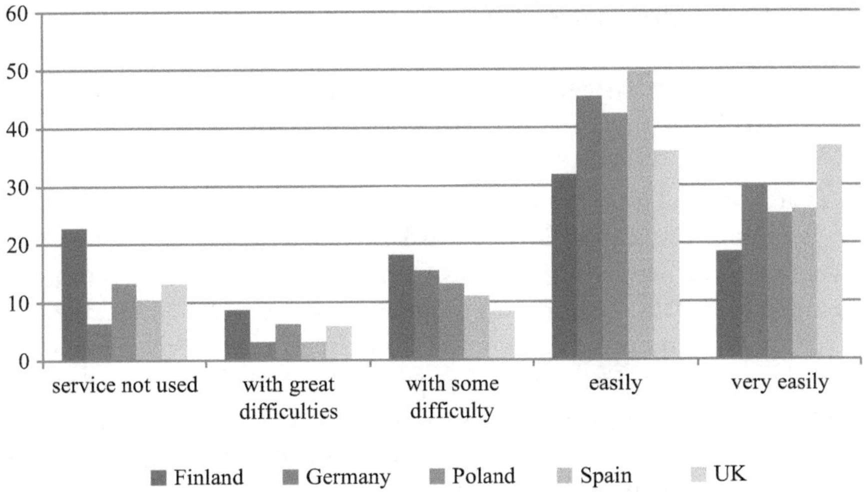

Fig. 8.2 How would you describe your access to public transport facilities, age group 65+. (Source: Eurofound 2012)

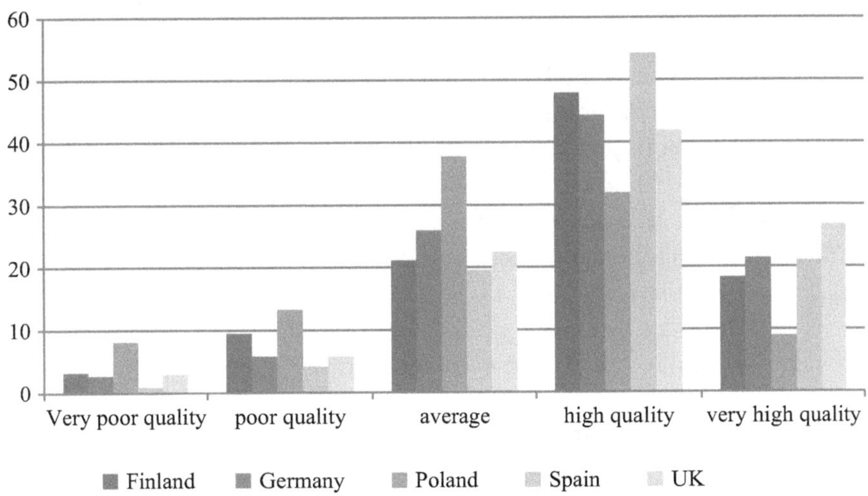

Fig. 8.3 How would you rate the quality of public transport? Age group 65+. (Source: Eurofound 2012)

The Deployment Gap

ICTs for Active Ageing as a Push Factor for the 'Silver Economy'

ICTs are a central aspect of Europe's economy and generate 25 per cent of the total business expenditure on R&D. Investments within this area account for 50 per cent of all European productivity growth (European Commission 2015). Due to the economic relevance, ICTs helped the idea of the 'Silver Economy' to gain momentum. In short, the 'Silver Economy' is a concept that sets its focus on the opportunities arising from the increasing longevity and the ageing of societies. The core idea of the approach is that there will be new and growing markets targeting people aged 50+. Due to the overall decline in population and the increasing number of older persons in relative and absolute terms, the consumer landscape is increasingly affected by the demands of older people. Companies and providers find new and growing markets for products and services designed for older customers and simultaneously have to face shrinking traditional markets. Moreover, the economic situation of older persons and their consumer behaviour have transformed dramatically. Along with demographic changes, the general socio-economic situation of older people in Europe has improved due to the development of pension schemes and other welfare features. Heinze et al. (2011) describe the Silver Economy as not a single but a cross-sectional market that comprises numerous economic areas such as health and care, mobility, housing, social and household services, tourism, education, financial services, areas of leisure, culture, new media, or ICTs. Especially the areas of housing, health, and social care supported by ICTs provide potential for all European member states (Domonkos et al. 2013):

> The Silver Economy is the new gold, Silicon Valley gurus tell us. Large new public and consumer markets of ICT products and services for ageing well are developing. Our ageing populations are drivers of economic growth. (European Commission 2015)

Despite this potential, the adoption, implementation, and diffusion of innovative technologies are still behind expectations. Europe is facing a deployment gap: huge research and development efforts on the one hand but only limited market success on the other.

However, each country has specific profiles and conditions, and large country differences have to be taken into account when targeting at cross-national cooperation. Ideas and solutions which are successful in one country may not be transferred simply to another country. Furthermore, with respect to active ageing, it must be underlined that people age differently and that economic, cultural, and regional characteristics have to be taken into account. In general, in CEE countries the technological state-of-the art is far less developed than in western European countries (although the need for socially innovative solutions is the highest). However, even in the 'early mover' countries the differences are enormous, partly mirroring how social risks and related needs of an ageing population have been both taken as point of departure for local social policies as well as incorporated in law making. This is particularly striking in Germany, where ICT-based measures reacting to the social risks of older people are far from being fully recognized as regular provisions in the German social insurance systems, which is noted for its poor dissemination (Henke and Troppens 2010). This is opposite to Finland, which our research identified as one of 'best practice'.

Overcoming the Deployment Gap

To address the deployment gap, our research concentrated on two issues: identifying the main barriers (Mollenkopf and Kaspar 2005; Mollenkopf et al. 2000) that negatively affect the dissemination of ICT-based products, services, and infrastructure for older persons, and finding ways to overcome them. To unfold its potential, technology needs to be adopted, implemented into different settings and—most importantly—used. There is an urgent need to influence the willingness to adopt technologies and the respective capabilities. In general, the acceptance of technology by older persons—and hence its use—is affected by numerous factors, not only by physical and cognitive changes accompanying the process of ageing, summarized on three different levels:

- The product (or innovation) itself.
- The user and his or her characteristics.
- Environmental and contextual aspects. (Merkel et al. 2016)

Because assistive technology is not only used by older persons themselves but also by their social support network as well as professionals (e.g. caring staff), it is a complex task to design technologies fitting into various contexts. Taking telecare and telehealth as examples, different aspects can be identified that affect the adoption, implementation, and use of technology, which can be grouped into technical, social, economic, and regulative dimensions, each covering multiple aspects. We reviewed 699 publications in the fields of telecare and telehealth which had been published in the EU between 2010 and 2015. The results reveal that barriers exist within five domains (see Table 8.1).

Figure 8.4 shows the distribution of these barriers according across different domains. The domain 'acceptance' was perceived as the most dominant, followed by financial and technical challenges. Factors that lead to non-acceptance—and non-use—cover multiple aspects. This includes safety concerns, for example, about data security, a lack of trust, for example, caused by poorly designed pilots, the general reluctance regarding telehealth and telecare, the feeling of stigmatization, generally low expectations, restrictions of autonomy, the fear of nursing staff being replaced by technology, a perceived increase of workload, the disruption of services and routines, the capability to understand the technology due to poor ICT skills or cognitive impairments, and so on (Merkel and Enste 2015). Even though the study focused on telecare and telehealth, most of the identified barriers can be transferred to other areas summarized under the label 'ICTs for ageing' including ambient/active assisted living. In particular, acceptance-related issues are relevant in all contexts where (modern) technology is used.

Using the domains and barriers as a basis, we screened for initiatives and projects that tried to address them and find ways in overcoming them. Table 8.2 shows selected examples of good practice in the field of housing, mobility, and health that were evaluated as part of our case-study research. The analysis of the cases underlines that different aspects of built and technological environments are closely linked to each other. ICT-based devices, services, and infrastructure are found in the context

Table 8.1 Category mapping for telecare and telehealth

Domains	Domains and determinants
Technical	Support Training Usability Personalization Interoperability Quality
Acceptance	Awareness Attitude Evidence based
Economical	Incentives Reimbursement Business models
Organizational	Organizational culture (e.g. leadership) External collaboration
Outer context	Standardization Security
Not specified	Transfer research into practice

Source: Merkel und Enste based on Broens et al. (2007) and Reginatto (2012)

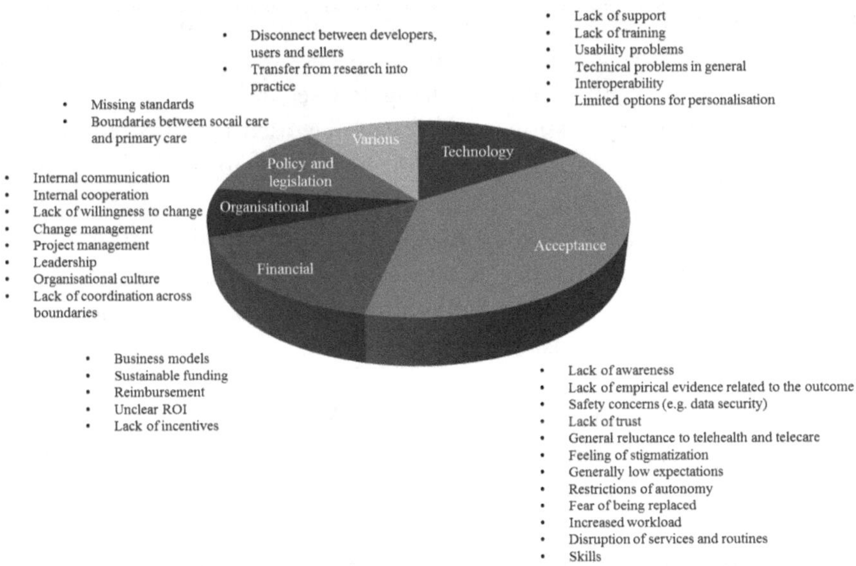

Fig. 8.4 Distribution of determinants. (Source: Merkel and Enste 2015)

of housing and mobility; this trend will continue in the near future. The case studies show that this asks for user-friendly approaches combining personal assistance, information policies, and trainings. The integration of volunteers—ideally peers—is identified as a concept or element that increases both the awareness and acceptance of new products and services as well as the chance of sustainability. Moreover, it supports the active participation of older people in society. User-integration is not only relevant at the beginning of the innovation process but also as a constant theme. Furthermore, the cooperation of multiple stakeholders at best as new strategic alliances is essential to develop and implement (advanced) approaches. In turn, this asks for measures to increase awareness of the users' needs but also of solutions that already exist.

The combined results of the literature review, secondary data analysis and the case studies, fed into a Delphi survey, to facilitate group consensus, and an international workshop with European stakeholders. The barrier that the respondents perceived as most dominant was the lack of awareness among older users as well as professionals. Interestingly, the misuse of personal data was seen as more important in terms of professional users and not primary end-users. Within the domain of acceptance, 'insufficient trainings for older users' was rated as the most important barrier; this changed during the second round and 'insufficient trainings for professional users' was regarded as more important. While legal issues were rated least important, problems with the technology (lack of interoperability, technology usability issues, missing technical standards, installation issues) received comparably high ratings.

In particular, 'good-practice' examples show that besides existing channels like websites or brochures, new forms of marketing are needed to increase awareness, such as:

- Within the good practice cases studied, Mi Liverpool has an innovative approach: a smart house was set up in the Museum of Liverpool. The smart house features several assistive devices that can be tested.
- Another good example was found in Finland where a 'mobile exhibition' was prepared containing a suitcase with several technical products that can be installed and tested temporarily within the living environment by interested users.

Table 8.2 Examples of good practice

Project	Short description
AAL-Lotsen ('AAL-guides')	The project aims at training older persons to become AAL-guides. The guides inform (older) people about the opportunities of AAL-solutions and show which products and services already exist
Andalusian telecare service	The Agency for Social Services and Dependency of Andalusia (ASSADA) provides social services for older and/or disabled persons including a telecare service
KÄKÄTE—User Centred Technology for Elderly People and Care Givers	The KÄKÄTE project seeks to find out more about using technology to support older persons in their daily routines and care givers in their assisting work. To reach these aims, the project combines research, networking and dissemination activities
Kutsuplus.fi Helsinki Regional Transport	Kutsuplus.fi combines public buses and taxis. Passengers can book a mini bus on short notice and chose a pick-up and drop-off point. The ride is shared if other passengers are going in the same direction at the same time
'Kotisairaala'—Hospital at Home	Kotisairaala provides intensive healthcare at the patients' home instead of a treatment in a hospital
Leicestershire County Council Public and Passenger Transport	Leicestershire County Council's aim is to provide transport services to their inhabitants by combining different services: Community transport, demand responsive transport and social car schemes
Mi—More independent	Mi seeks to increase the scale of technology-enabled services operating in Liverpool. The initiative combines several ways to raise awareness on products. This includes, for example, a webshop or an exhibition in the local museum
Nie Sami ('Not Alone')	Nie Sami is social housing project for elderly, offering assisted housing in rented flats
SehrMobil Siegen ('VeryMobile Siegen')	SehrMobil is a platform that offers a variety of age-friendly mobility options and also provides information about events and connects people
Silver Internet	Silver Internet encourages older persons to get in touch with modern technologies. Under professional supervision, people can help each other and discuss problems related to the use of ICT-based devices

- A third interesting approach found, for example, in the United Kingdom (Mi Liverpool) and Germany (AAL-Pilots VdK Saarland) refers to programmes aiming at user empowerment. Older persons are trained to experts for assistive technologies and inform their peers about available solutions and help them with the set-up and maintenance (Lang 2018).
- A forth 'good practice' example comes from Spain, belonging to the action field of mobility: The vertical transport system Donostia-San Sebastian links high and low districts of the city with an elevator, escalator or ramp, using technology for making walking and cycling trips to and from the city centre easier for citizens living in the hilly party of the city. This system was installed with the active participation of the older residents.

Such examples show new approaches that can easily be copied and integrated in new projects and initiatives. Moreover, solutions to overcome the lack of awareness were seen in building on local and regional initiatives driven by local and regional stakeholders. Workshops and the training of older users, as well as professional users were seen as essential to promote technology and to overcome the 'lack of awareness'. Peer support—older persons helping each other with new technologies or 'scouts' identifying barriers in public transport systems and scandalizing them (Lang 2018)—was rated as a promising approach and should be focused on.

The lack of standardization and interoperability was seen as very important in particular in the area of healthcare. Therefore, this issue has to be tackled first when it comes to scaling up telecare and telehealth services—not only for older persons. However, it was felt that technology alone should not be the centre of attention, rather the deficits in the professional care should be the starting point for intervention, integrating older persons affected, and with need for social assistance into a local care system with technological assistance.

An increase in public funding was mentioned as a way to tackle the deployment gap. A way to secure the funding of new products, services, and infrastructure was seen as venture capital—though this was not rated very highly. The question of how, and by which stakeholder the respective financial provisions guaranteeing a long-term utilization, are financed and do not overburden the end-users financially, is far from being solved in the countries we studied. A few experts saw the main responsibility on the side of the national social security schemes as a necessary public reaction to demographically caused new social risks and needs. If private investors find the market promising, they will start to invest money in it. However, at this time, a common European market is missing and is seems unclear if there will be one in the near future.

- A further good example for both better using technologies to prolong independent living as well as to promote a silver housing economy comes from Germany, where different modules of smart living are offered (cf. GdW/InWIS/SIBIS 2015). In the field of prevention, home automation also appears to be a growing market. If e-health and telemedicine applications can both increase the quality of treatment and lower the costs, the question arises why the available findings have not yet been taken up across the board and been put to practice in standard care. Meanwhile, the pilot phase has been more than completed and the housing industry as well as the tenants are ready for practical implementations. First isolated projects have already demonstrated the positive effects of home networking.
- Another 'success story' from the 'Silver Economy Initiative' in Germany was the promotion of barrier-free housing solutions (e.g. bathrooms). Craftsmen were trained to advise customers and were sensitized to the need of ageing customers. This could also be a promising starting point to promote ICT-based technologies for independent living. There need to be places and people who can be addressed with technical questions. While initiatives that rely on volunteers (e.g. AAL-guides) are a good way to foster user engagement, learning processes, as well as inter- and intragenerational approaches, maintenance and training need to be carried out by professionals.

There are, however, various barriers to be overcome. Empirical experience shows that the new technological solutions are often perceived as impersonal and technocratic. Therefore, the technology-supported value-added services must not be too technology-centred but also have to take into account social factors. Furthermore, surveys show that the willingness to pay for home-based services is still relatively low. These considerations can explain why most of the concepts of smart living or ambient assisted living have not yet overcome the status of research and development projects and have not been adopted in standard care so far. Up to now, the necessary interfaces with the potential areas of application only work in selected projects but the application is slowly spreading. Above all, the technological development must focus on the user benefits (added value).

One further prominent barrier was the mismatch of the real user needs, and the images, which designers and developers have of older persons. Although there has been extensive research on needs of older users and how technology can help them both to live independently, as well as increase their quality of life, older persons still are seen by many designers and developers as passively responding to their environments (Peine and Neven 2011). To address the issue of non-acceptance and non-use, participatory development and design approaches have proven to be promising. Research has shown that users actively engage with technology and modify it (Wherton et al. 2015). They shift the centre of the design process to the users, their needs, and characteristics. The field of participatory approaches is very inter- and transdisciplinary covering psychology, social sciences, neurosciences, linguistics, as well as technical disciplines. This makes it hard, if not impossible, to provide a complete overview. For instance, the AAL-Toolbox (AAL Association 2013) offers multiple methods that can be used to integrate the target group during the conceptualization, testing, or pilot phases.

The design processes often tend to concentrate on the user's needs. However, focusing the design solely on previously defined needs bears the risk of excluding the users. Design methods have to consider the whole process ranging from developing the concept, refining it to creating a prototype and field trials. To fully understand how technology will be used (by older persons), design processes have to see how products and services develop in the 'real world'—the user's hands. This cannot be achieved by simply asking them about their expectations and needs. Birken et al. (2016) argue that most of the tools and methods aim at testing a prototype or mock-up, but rarely pay attention to the needs of potential users that derive from challenges occurring during their daily tasks. Hence, the authors propose to address this issue and evaluate the demands of the users including their patterns of lifestyle and habits within in their home environment, based on the concept of *orientation to the real living conditions* (Lebensweltorientierung) of the people concerned. Similar methods have been described previously; such as ethnographical studies on the implementation of assistive technologies in living arrangements of older persons (Wherton and Monk 2008; Greenhalgh et al. 2016).

Although the evaluation of cost-effectiveness was not perceived as a major barrier, it was suggested that we need to rethink evaluation design. The evaluation of the evidence base should answer which applications work, in which specific contexts, and for which target populations. It was argued that a discussion is needed on what level of evidence is feasible and necessary for different interventions. It was proposed that frameworks for the evaluation of projects, products, and initiatives are needed and should cover the economic, technical, legal, and social dimensions. The evaluation of technologies covers several dimensions: the individuals and their adoption and use, the organizational context, as well as the broader context. Greenhalgh et al. (2016: 1) argue that research to date into assisted living technologies broadly consists of three generations: technical design, experimental trials, and qualitative studies of the patient experience. They propose to pay attention to a fourth generation of studies which see technology as a dynamic, networked, and potentially unstable system.

Our research also had a special focus on mobility and transport—topics on the rise in ageing societies, but very often underestimated in the public discourse of adapting infrastructural environment to the (often changed) needs of older persons. However, mobility is also important to facilitate a self-determined life and social inclusion and is a key factor for life satisfaction in old age. The demographic change implies many challenges for public transport—above all an expected/assured higher share of people with impairments and a growing urban-rural divide. One barrier for the usage of public transport is low availability, foremost in rural areas. In Europe, there are several regions that are difficult to reach. Here, people with reduced mobility are specifically dependent on flexible transportation arrangements, at best organized as seamless transition between different options which demands a local mobility management including service delivery (e.g. shopping) (BMFSFJ 2016). As seen in the national policies of the target countries and in the European policies, accessibility is the major concern at all levels. Accessible vehicles and stations are very important for people with limited mobility. Moreover, public transport of the future has to take into account the heterogeneity of the people which is increasing with age. Not only do older persons differ widely because of various ways of living, but also because of the range of mental and physical impairments. Respective diverse needs and requirements of older people have to be considered when developing measures and offers for mobility and transport. A strong driver for mobility is the technical development that helps older people to orientate and inform themselves before and during the trip for seamless door-to-door transportation. Another driver for mobility policies is the aim to decrease CO^2 emissions through a higher usage of public transport instead of car driving.

Implications for Policy and Practice

The challenges to adopt, implement, and disseminate products, services, and infrastructure are numerous and complex and a generalization or simplification would neglect contextual influences. Tackling the existing barriers requires a keen understanding of national political and societal contexts. However, there are overarching aspects and common challenges that need to be addressed Europe-wide. The analysis underlines that the different aspects of built and technological environments are closely linked to each other. ICT-based devices and services are found in the context of housing and mobility; this trend will continue in the near future. Figure 8.5 shows the combined results of the different steps undertaken within MoPAct, including the most dominant barriers that hinder the uptake of ICT-based products and services for older persons and potential solutions.

Avoid a Technology Divide Special attention has to be paid to the concept of the Silver Economy and the importance of social innovations. Many

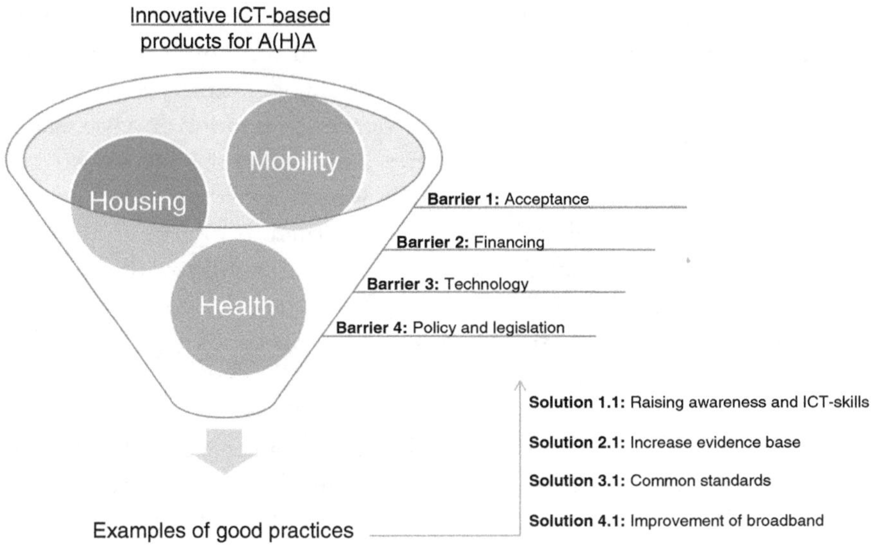

Fig. 8.5 The deployment gap and potential solutions

innovative products and services are technology-driven. Against the background of the digital divide, an increasing uptake and use of ICT-based products bears the risk of an unequal distribution among members of society. If this occurs regularly between the same social groups, it could be perceived as inequity and hence become a social problem (Hoffmann 2008 based on Hradil 2001). A politically promoted strategy to increase the uptake of ICT-based products and services is in danger of benefiting only parts of the society and could lead to new social inequalities. R&D projects should take into account the heterogeneity of users. Those who participate in projects or act as volunteers have the financial, cognitive, and societal resources in most cases.

Raise Awareness The lack of awareness was perceived as the most dominant barrier. New products and services, projects, and initiatives need to address marketing aspects. This includes not only older persons as primary users, but also secondary users like relatives and professional care staff. Projects need not only to include approaches to increase marketing, but also larger strategies are needed to inform about what products and services are available and how these can support active and healthy ageing.

Foster Participatory Design *User involvement* is a key issue. The user's needs and characteristics have to be considered during all stages of the design process. Co-design and co-production of products and services helps to raise awareness and leads to an increased adoption and use. More research and funding could help to better understand the user's need and how they change over time. Though it was shown that older persons experiment with technology and modify it, there is often only limited space for alterations. From the very beginning of the innovation process products should be designed in a way that users are able to interact with the product. Too extensive research on user's needs could negatively affect the engagement with the finalized product of older persons.

Increase of ICT-Skills The lack of ICT-literacy is a central challenge that needs to be addressed. In particular, older users often have only limited ICT-skills. ICT-literacy has to be increased including all age groups. The fear of misuse of personal data could be reduced and the willingness to actively engage with modern technologies increased. This includes primary as well as secondary users. The implementation of new products and services requires strategies to introduce the technology to the users and has to be adopted according to their changing skills, needs, and competences. Training should not concentrate on a single point in time and instead be thought of as a process. Older persons, as well as professional users, need a contact person when questions arise, who can help with the product's maintenance.

Improve Broadband and Mobile Internet Connections More and more products and services for older persons rely on ICT; this includes areas like health and social care, housing, mobility, and leisure activities. To cover all citizens, Internet accessibility has to be improved especially in rural areas.

Define Common Standards Interoperability is a major barrier. Many systems are not compatible with technology in place, like the telephone at patient's homes. Moreover, interoperability is necessary for the delivery of health and social care when multiple partners at different levels are involved. To increase the interoperability of technical solutions, common

European standards are needed. In particular within healthcare, this has been neglected for too long. Although European standards like IHE6 address this issue, the debate has to be expanded and also address AAL or assistive devices in general.

Increase of Sustainable Funding The sustainability of publicly funded projects and initiatives is insufficient. Funding programmes increasingly focus on such aspects, especially those that support the development of products. However, many initiatives struggle shortly after the funding has ended. At best, ICT support for persons in need should be implemented as a standard provision within the national social security systems. The success of the silver market depends on a durable demand being secured by a sustainable funding. This corresponds with a more general experience: The dissemination of those social services being recognized as merit goods aiming at an equitable distribution of public goods need a solid financing without incorporated social barriers (Heinze et al. 2011).

Identification and Dissemination of Business Cases Successful business cases are still missing—especially in terms of social business cases. The combination of technical and social innovations offers many chances in this respect. Promising examples need to be disseminated and adopted to specific contexts.

Strengthen the Household as a Health Location In connection with the spread of new telehealth and telecare devices that facilitate an individualized treatment (or rehabilitation) at home, the issue of the household as the 'third' health location is generally being broached of lately. However, although there are some best practice experiences, it is not being implemented as a matter of course yet. One has to keep in mind though that despite all euphoric appraisals, telehealth and telecare depend on environments that are often not well-integrated, for example, due to segmentation. It is therefore difficult to break down the prevalent sectoral division between inpatient and outpatient treatment and to introduce innovative care models. These goals can only be achieved step-by step. This is also true for the relationship between social service organizations/nursing services and the housing industry which have only begun to confer with each other in the past few years. Moreover, many discussions and policy recommendations are adversely affected by the fact that the focus is too strongly on the housing space instead of an integration of the local living environment and its inhabitants into the restructuring process.

Further Research Priorities

There has been extensive research on reasons for the non-adoption of modern ICT-based technology by older persons. Still, to fully unlock the potential that technology offers for the process of active and healthy ageing, several aspects need to be better understood. The research within MoPAct has revealed multiple themes that need to be investigated in future research. We want to highlight in particular three aspects:

Social Inequalities and Technology Whether or not a person is able to afford and use technology depends largely on his or her socio-economic background, which includes education, income, gender, and household size. An increasing proliferation of ICT-based products and services therefore bears the risk to foster social inequalities. To overcome this problem, research needs to concentrate on those who could actually benefit most of assistive technology: people who are living alone without a social network that supports them. At the moment, a lot of research results are biased because those participants who engage in studies on new products and services often have a high socio-economic status and are more likely able to buy and use modern devices. It would therefore help to address users who typically are under the radar. Against the background of the Silver Economy, a concentration on the economic factor age could increase inequalities due to its focus on the privileged elderly with a high spending power. Though there is evidence that tomorrow's older persons are more open to ICTs, it can be assumed that, due to cohort effects, there will still be the comparable challenges in the future. Challenges that have been described already ten or more years ago are still eminent (Mollenkopf 2016).

User Involvement and Co-production Participatory research methods are seen as a key aspect within gerontechnological innovation processes; user engagement is seen as a necessary way to increase the acceptance of new products (and services). However, the field of participatory research is broad and interdisciplinary. Different concepts have emerged, including co-production, design thinking, and user experience. Within the context of gerontechnology, this means, for example, to have senior citizens participating within the project not only as untrained designers but also, of course, as users. While the basic idea of including users and other actors into the research and design process is a common theme, an approach addressed in particular to older users is still missing. There have been some initiatives lately showing good examples (e.g. Greenhalgh et al. 2016) but these have not been transferred into practice.

Larger Studies on the Socio-cultural Background Though there are several larger studies available taking into account the many facets of technology adoption, implementation, and diffusion, these usually focus on single countries. What is missing is a European perspective comparing different countries. While databases like Eurostat or the European Quality of Life Surveys offer good starting points, the scope needs to be extended.

Conclusions

There are many promising ICTs-based products and services that have the potential to support active ageing and, moreover, have the potential to generate economic growth. However, Europe is facing a deployment gap. ICTs are a strong instrument for making public infrastructure, services, and products more age friendly and the work undertaken so far shows great untapped potential. The reasons for this are complex and include multiple barriers that can be grouped into the multiple domains. Currently, the most dominant challenges are the lack of acceptance. This includes, in particular, the lack of awareness among potential users which leads to non-adoption and non-use of products and services. Solutions are needed that not only raise awareness on new products and—even more important—on products that are already available on the market. Technical solutions will only succeed and be accepted by the market if a significant added value for the user is created, which can only lie in the quality of life. Many projects are still technology-driven, although just recently social innovations have emerged that offer the potential to create this added value. The examples of good practice that we have studied in MoPAct have underlined the potential that exists within this context. To unfold this potential and ensure sustainability, new funding strategies and social business cases are of utmost importance. Therefore, a combined approach of technical and social innovation should be encouraged. While differences among the European countries in terms of ICTs-uptake and literacy/skills can be a chance for policy learning and knowledge transfer, more country-specific approaches would be beneficial.

Each of the three areas—ICTs, housing and mobility—offers possibilities to support independent living and to improve the quality of life of older persons. It should have become clear that all three topics are closely linked to each other and therefore should be thought together. However, this raises the complexity. While the dominant driver of innovative solutions within all three areas can be found in demographic change, each topic has its own challenges. With ICTs, the dominant problem can be seen in the lack of diffusion. Concerning housing, one of the biggest problems is the financing of the necessary adaptions of the dwellings, the environments, and different initiatives in the neighbourhoods. Moreover, innovative homecare solutions ask for integrated solutions that pass the borders of different policy sectors (e.g. housing, health, and care).

Taking a closer look at those barriers, it becomes clear that the countries in Europe are facing common challenges. Nevertheless, each country has special characteristics and conditions that have to be taken into account. Especially when targeting at cross-national cooperation this has to be kept especially in mind. Ideas, projects, or solutions which are successful in one country may not be transferred to another without modifications. Furthermore, with respect to active and healthy ageing, it must be underlined that people age differently. These insights may not be new; still, there seems to be a lack of awareness.

The central question to be posed is: To what extent can ageing societies benefit from ICTs-based housing and mobility solutions? Of course, there is no single answer to this question, but a plethora of possible approaches exists. As shown, a central challenge can be seen in the transfer of scientific knowledge. Much effort is put on R&D, but only a fraction of the results are successfully implemented and diffused. Therefore, this implies to focus more on the introduction to the market.

Another approach aims at rethinking the innovation system. This refers to an approach developed by the High Level Group on Innovation Policy Management, which postulates a

> radical change in innovation policy seems indispensable, from fragmentation to coordination, from a narrow science and technology orientation to an all-encompassing, holistic and coherent strategy involving several policy areas, from a diffuse to a highly focussed division of labour between all the players and stakeholders involved. (HLGIPM 2013: 9)

This approach includes seven key activities:

- Optimize the embryonic European innovation ecosystem
- Improve policy coherence
- Reduce regulatory complexity and rigidity
- Eliminate obstacles and provide new funding to innovations
- Facilitate industrial cooperation and re-interpretation of competition law
- Take an encompassing and inclusive view of intellectual property
- Increase the innovation potential through user and consumer drive

From the perspective of an ageing European society, a central focus could be achieved by drawing the attention especially to the last bullet point. User-integration, a design for all, and the understanding of the targeted population can be regarded as essential. But this also means learning from failed examples to accepting that some products and services are just not suitable or address needs which are already covered by existing solutions.

References

AAL Association. (2013). *AAL*. Retrieved March 20, 2018, from AAL-Toolbox. http://www.aal-europe.eu/wp-content/uploads/2015/02/AALA_ToolboxA5_online.pd

Birken, T., Pelizäus-Hoffmeister, H., & Schweiger, P. (2016). Technische Assistenzsysteme und ihre Konkurrenten: Zur Bedeutung von Praktiken der Alltagsbewältigung für die Technikentwicklung. *Zukunft Lebensräume – Kongress 2*. Frankfurt/Main.

BMFSFJ. (2016). Bundesministerium für Familie, Senioren, Frauen und Jugend (Hrsg.). *Siebter Altenbericht. Sorge und Mitverantwortung in der Kommune*. Berlin: Aufbau und Sicherung zukunftsträchtiger Gemeinschaften.

Bouma, H. (2009). *Gerontechnology for Serving Needs and Ambitions of Older Adults*. Retrieved March 20, 2018, from Innomed. http://www.innomed.no/media/uploads/samhandling_velferdsteknologi_og_innovasjon/Herman_Bouma.pdf

Broens, T., Veld, R., Volenbroek-Hutten, M., Hermens, H., van Halteren, A. T., & Nieuwenhuis, L. J. M. (2007). Determinants of Successful Telemedicine Implementations: A Literature Study. *Journal of Telemedicine Telecare, 13*, 303–309.

DEFRA. (2004). *Department for Environment, Food and Rural Affairs: Review of the Rural White Paper Our countryside: The Future*. DEFRA. https://www.thenbs.com/PublicationIndex/documents/details?DocId=265877

DETR. (2000). *Department of the Environment, Transport and the Regions: Our Countryside: The Future. A Fair Deal for Rural England*. Rural White Paper. DETR.

DfT. (2012a). *Department for Transport: Making Transport More Accessible to All*. London. Retrieved from https://www.gov.uk/government/policies/making-transport-more-accessible-to-all

DfT. (2012b). *Department for Transport: Green Light for Better Buses*. London. Retrieved from https://www.gov.uk/government/uploads/system/uploads/attachment_data/file/3617/green-light-for-buses.pdf

DfT. (2012c). *Transport Solutions for Older People. Information Resource for Local Authorities*. London: Department for Transport. Retrieved from https://www.gov.uk/government/uploads/system/uploads/attachment_data/file/4478/transport-solutions-for-older-people.pdf

Domonkos, T., Horvat, P., Hvozidikova, V., Lichner, I., Miklosovic, T., & Palenik, V. (2013). *Modelling the Economic Potential of the Silver Economy*.

NEUJOBS Working Paper, 12.3. Retrieved from http://www.neujobs.eu/sites/default/files/NEUJOBS%20Working%20Paper_Modelling%20Silver%20Economy_12.3.pdf

Ehlers, A., & Naegele, G. (2017). Soziale Ungleichheit und digitale Inklusion – ein relevantes Thema auch im Alter. In Generali Deutschland AG (Hrsg.). *Generali Altersstudie 2017. Wie ältere Menschen in Deutschland denken und leben* (pp. 119–122). Berlin: Springer.

Eurofound. (2012). *Third European Quality of Life Survey – Quality of Life in Europe: Impacts of the Crisis*. Luxembourg: Publications Office of the European Union.

European Commission. (2015). *Growing the European Silver Economy.* European Commission. Retrieved from http://ec.europa.eu/research/innovation-union/pdf/active-healthy-ageing/silvereco.pdf

Fiedler, M. (2007). *Older People and Public Transport. Challenges and Chances of an Ageing Society.* Cologne.

Follmer, R., Gruschwitz, D., Jesske, B., Quandt, S., Lenz, B., Nobis, C., … Mehlin, M. (2010). *Mobilität in Deutschland 2008.* Bonn: Ergebnisbericht. Struktur – Aufkommen – Emissionen – Trends. INFAS Sozialforschung.

Gast, R. (2013). Der unsichtbare Pfleger. Ingenieure und Informatiker haben einen Traum: Hightech soll im Alltag über die Gesundheit alter Menschen wachen. *Es wird geforscht, getestet und nach Geldgebern gesucht, 11.*

GdW, InWIS, & SIBIS. (2015). *Technische Assistenzsysteme für ältere Menschen – eine Zukunftsstrategie für die Bau- und Wohnungswirtschaft.* Berlin: GdW.

Greenhalgh, T., Jackson, C., Shaw, S., & Janamian, T. (2016). Achieving Research Impact Through Co-creation in Community-Based Health Services: Literature Review and Case Study. *The Milbank Quartley*, 392–429. https://doi.org/10.1111/1468-0009.12197.

Heinze, R. G. (2015). Wohn- und Lebensformen im Alter: Selbstständigkeit erhalten. Integration fördern Der Bürger im Staat, 2–3.

Heinze, R. G. (2016). Digitalisierung und Wohnen: das vernetzte Zuhause und individualisierte Technologien als Option für das Wohnen im Alter. *Zeitschrift für Sozialreform, 64*(4), 443–471.

Heinze, R., & Naegele, G. (2013). Social Innovations in Ageing Societies. In H. W. Frantz, J. Hochgerner, & J. Howaldt (Eds.), *Challenge Social Innovations* (pp. 153–168). Wiesbaden: Springer.

Heinze, R. G., Naegele, G., & Schneiders, K. (2011). *Wirtschaftliche Potenziale des Alters.* Stuttgart: Kohlhammer.

Henke, K.-D., & Troppens, S. (2010). Zur Finanzierung assistierender Technologien. In U. Fachinger & K.-D. Henke (Eds.), *Der private Haushalt als Gesundheitsstandort Theoretische und empirische Analysen* (pp. 135–146). *Baden-Baden: Nomos-Verlag.*

Highlevel Group on Innovation Policy Management (HLGIPM). (2013). Report and Recommendations. Online: https://docs.wixstatic.com/ugd/4bc46b_35ca1b1f47514d2a85c794f8be5ebeb2.pdf

Hoffmann, R. (2008). *Socioeconomic Differences in Old Age Mortality*. Dordrecht: Springer.

Hradil, S. (2001). *Soziale Ungleichheit in Deutschland. Lehrbuch*. Wiesbaden: Springer VS.

Lang, A. (2018). *Herausforderung erkannt – Sozialverband VdK unterstützt Mitglieder und Bürger beim Einsatz technischer Hilfen im Alltag. Vortragsmanuskript*. Berlin: Fachkonferenz „Kommunen in der alternden Gesellschaft".

Lupton, D. (2016). *The Quantified Self. A Sociology of Self-Tracking*. Cambridge: Polity.

MEC. (2013). Ministerio de Economía y Competitividad (2013). Spain: Estrategia Española de Ciencia y Tecnología y de Innovación 2013–2020.

Merkel, S., & Enste, P. (2015). *Barriers to the Diffusion of Telecare and Telehealth in the EU: A Literature Review*. Institution of Engineering and Technology: IET International Conference on Technologies for Active and Assisted Living (TechAAL), London, p. 6.

Merkel, S., Enste, P., Hilbert, J., Chen, K., Chan, A., & Kwon, S. (2016). Technology Acceptance and Aging. In S. Kwon (Ed.), *Gerontechnology. Research, Practice, and Principles in the Field of Technology and Aging*. New York: Springer.

Ministry of Infrastructure. (2005). *National Transport Policy for 2006–2025*. Poland.

Mollenkopf, H. (2016). Societal Aspects and Individual Preconditions of Technological Development. *Geron, 15*(4), 216–226. https://doi.org/10.4017/gt.2016.15.4.011.00.

Mollenkopf, H., & Kaspar, R. (2005). Elderly People's Use and Acceptance of Information and Communication Technologies. In B. Jaeger (Eds.), *Young Technologies in Old Hands – An Unternatuional View on Senior Citizens' Utilization of ICT* (pp. 41–58). Copenhagen: DJOF Publishing.

Mollenkopf, H., Meyer, S., Schulze, E., Wurm, S., & Friesdorf, W. (2000). Technik im Haushalt zur Unterstützung einer selbstbestimmten Lebensführung im Alter [Everyday Technologies for Senior Households]. *Zeitschrift für Gerontologie und Geriatrie, 33*(3), 155–168.

MTC. (2008). Ministry of Transport and Communications: Transport Policy Guidelines and Transport Network Investment and Financing Programme Until 2020 – Government Transport Policy Report to Parliament. MTC.

Naegele, G., & Bauknecht, J. (2017). *Final Report for WP3 ("Employment"): Strategies for Extending Working Lives, Raising Older Workers Employment Rates and Intensifying Lifelong Learning.* Dortmund: Forthcoming.

Oswald, F., & Wahl, H. -W. (2016). Alte und neue Umwelten des Alterns – Zur Bedeutung von Wohnen und Technologie für Teilhabe in der späten Lebensphase. In E. Naegele, A. Olbermann, & A. Kuhlmnn (Eds.), *Teilhabe im Alter gestalten.* Wiesbaden: Sprigner VS.

Peine, A., & Neven, L. (2011). Social-Structural Lag Revisited. *Geron, 10*(3). https://doi.org/10.4017/gt.2011.10.3.002.00.

Reginatto, B. (2012). Understanding Barriers to Wider Telehealth Adoption in the Home Environment of Older People: An Exploratory Study in the Irish Context. *International Journal on Advances in Life Sciences, 4*(3&4), 63–76.

Selke, S. (2016). *Lifelogging. Digitale Selbstvermessung und Lebensprotokollierung zwischen disruptiver Technologie und kulturellem Wandel.* Wiesbaden: Springer VS.

van Bronswijk, J., Bouma, H., Fozard, J. L., Kearns, W. D., Davison, G. C., & Tuan, P.-C. (2009). Defining Gerontechnology for R&D Purposes. *Geron, 1.* https://doi.org/10.4017/gt.2009.08.01.002.00.

van den Berg, N., Schmidt, U., Stentzel, H., Mühlan, W., & Hoffman, R. (2015). Telemedizinische Versorgungskonzepte in der regionalen Versorgung ländlicher Gebiete. Möglichkeiten, Einschränkungen, Perspektiven. *Bundesgesundheitsblatt.*

Webber, S. C., Porter, M., & Menec, V. H. (2010). Mobility in Older Adults: A Comprehensive Framework. *The Gerontologist, 50,* 443–450.

Wherton, J. P., & Monk, A. F. (2008). Technological Opportunities for Supporting People with Dementia Who Are Living at Home. *International Journal of Human-Computer Studies, 66*(8), 571–586. https://doi.org/10.1016/j.ijhcs.2008.03.001.

Wherton, J., Sugarhood, P., Procter, R., Hinder, S., & Greenhalgh, T. (2015). Co-production in practice. How people with assisted living needs can help design and evolve technologies and services. *Implementation Science, 10*(75). https://doi.org/10.1186/s13012-015-0271-8.

9

Social Support and Long-Term Care for Older People: The Potential for Social Innovation and Active Ageing

Katharine Schulmann, Monika Reichert, and Kai Leichsenring

Introduction

As noted in Chap. 1, the terms 'active ageing' and 'social innovation' have recently gained special attention in different policy fields (see e.g. European Innovation Partnership on Active and Healthy Ageing; Howaldt et al. 2016; Walker and Foster 2013) because they are seen as important ways to overcome the challenges that go along with demographic and social change in Europe. One of these challenges is the rising number of people in need of long-term care (LTC) and social support, specifically the matching between supply of and demand for social

With contributions from (in alphabetical order): G. Casanova, M. Ghenta, G. Lamura, A. Matei, I. Mladen, G. Naegele, G. Paat-Ahi, S. Schulze, Z. Széman, M. Tróbert.

K. Schulmann (✉) • K. Leichsenring
European Centre for Social Welfare Policy and Research, Vienna, Austria
e-mail: schulmann@euro.centre.org; leichsenring@euro.centre.org

M. Reichert
TU Dortmund University, Dortmund, Germany
e-mail: Monika.Reichert@fk12.tu-dortmund.de

A. Walker (ed.), *The Future of Ageing in Europe*,
https://doi.org/10.1007/978-981-13-1417-9_9

support and LTC. However, neither the rhetoric of active ageing nor the current discourse on social innovation is considering these areas as a major source for new ideas and positive change in the social construction of old age. To make longevity a real asset for society, it is therefore necessary to change current socio-economic perceptions and approaches to care as well as existing organisational and governance structures, in order to achieve a more active and inclusive understanding and experience of the ageing process, also and especially in case of LTC needs.

In the context of MOPACT our work on social support and LTC for older people paid special attention to this matter by describing, comparing, and analysing different approaches to fund, organise, and regulate LTC in eight European countries (Austria, Estonia, Finland, Germany, Hungary, Italy, Romania, and Portugal).

The main objective of this research was to identify potentials for and ongoing practices of social innovation and active ageing in the area of LTC[1] especially in the eight aforementioned countries. This included identifying the key factors that drive innovation in social support and LTC, innovation that then, in turn, can aid the transformation of this sector into an engine of sustainable socio-economic progress. The identification of these key factors was the basis for building comprehensive future scenarios describing how integrated, age-friendly approaches can be realistically implemented to ensure active ageing via inclusive social support and LTC practices across diverse European welfare states. This research serves to consolidate knowledge and inform social support and LTC policies at European Union (EU) and national levels, a general need that has only recently been further emphasised (Deusdad et al. 2016, p. 144).

This chapter is divided into the following sections: first, key challenges with regard to LTC in Europe are presented with a special emphasis on the concept of care regimes. Second, the concepts of active ageing and social innovation and their relevance to the field of LTC are discussed. The third and main part of this chapter is dedicated to the results of the research carried out by research teams from the eight countries. Starting from the identification of examples of social innovation, the key drivers (and barriers) of innovative programmes and practices in LTC are analysed, and then those areas with the greatest potential and need for innovation are discussed. A short summary and some suggestions for further research priorities conclude this chapter.

The Key Challenges with Regard to LTC in Europe

Socio-Demographic Developments and Their Consequences for LTC

The countries of Europe are ageing, yet at a different pace and with huge differences in political, cultural, social and economic framework conditions. Across the EU, 93.9 million people are aged 65 and over, equating to 18.5% of the total population (ENNHRI 2017). Although the projected changes in population structures vary considerably among the individual EU Member States, the share of older people in the total population of the EU is projected to increase to 28.7% (or 149.1 million elderly persons) by 2080. Europop2013 projections also indicate that there will be substantial increases in the number of the older old—defined here as those aged 80 years or over. Their share in the total EU population is projected to increase from 5.1% (26.0 million) in 2014 to 12.3% (63.9 million) by 2080 (Eurostat 2015). The reasons for this demographic shift are well known: decreasing fertility rates combined with rising longevity resulting in what is nowadays commonly called 'ageing societies'.

Most individuals older than 65 years are relatively healthy and live independent lives. However, according to the latest data, about 45% are suffering to a greater or lesser degree from chronic and often multiple diseases, with physical, mental and/or cognitive impairments, and between 20% and 33% of these people require some kind of support and care (ENNHRI 2017). This is especially true for individuals 80 years and older because very old age is linked with the risk of dependency. In 2011 the OECD reported that 50% of people receiving LTC were 80 years and older (Colombo and Mercier 2011). Provided there is no breakthrough in combating chronic diseases like arthritis or dementia, the need for LTC will rise exponentially in the near future. LTC is defined as '.... the provision of informal, formal and unregulated care assistance to older persons by family members, public, private and not-for-profit care services and migrants' (Anderson 2012, p. 135).

Besides demographic developments, there are other factors that contribute to LTC challenges. Although the provision of LTC is complex and divided between state, market and family care to a greater or lesser degree in different countries (Timonen and Doyle 2007), strengthening informal care—traditionally carried out by a female family member—is a policy aim in many EU Member States. However, this aim is in conflict with the fact that more and more women have joined the labour market. This positive development is in accordance with the Lisbon Treaty and the EU 2020 Strategy to increase the (female) labour force participation but it has also led to the emergence of 'new risks' throughout the life cycle, such as the difficulty in reconciling work and family responsibilities (Hoff et al. 2014; Reichert 2012). A decrease in the informal care potential is caused by other factors as well. Especially Mediterranean but also many Central and Eastern European (CEE) countries, most of which have based the provision of LTC on the significant contributions of women and other family members, are being confronted with declining birth rates, new family formations, and structural changes in the female role. One consequence of these developments is that a growing number of (older) persons in need of LTC live alone and/or are childless and therefore cannot count on the help of close family members. Another pressure refers to the LTC workforce. There is a current workforce shortage in many EU Member States with the result that projected future demands for LTC will not be met. Until now, the recruitment and retention of professional (nursing) staff, who could 'substitute' for deficits in the pool of informal carers in the community and who could provide LTC in institutions, has proven difficult (Greve 2016). As LTC is labour-intensive, countries' financial capacity to adequately respond to these social changes and needs is decreasing in the aftermath of the 2008 economic and financial crisis, which has been almost exclusively addressed by policies of austerity. However, against the backdrop of the described trends, experts expect 'Public LTC expenditure … to at least double and possibly triple by 2050' (Colombo et al. 2011, p. 73). With this projection in mind, countries in the EU are being forced to find new and innovative ways to finance LTC as well as to re-organise and re-distribute care responsibilities within societies (Lutz and Palenga-Mollenbeck 2010). Due to the large diversity among EU countries in the way LTC delivery has been shaped and funded, it seems that some countries are better prepared for this challenge than others. The following section will shed some light on the role of LTC in different welfare regimes.

The Heterogeneity of LTC Systems in Europe: The Role of Care-Regimes

The development of LTC delivery and the extent to which LTC has been acknowledged as a social risk can only partly be explained by the traditional categories of welfare state regimes that delineated liberal, conservative/corporatist and social-democratic welfare states (Esping-Andersen 1990). Indeed, Esping-Andersen's categorisation has been criticised not only for its emphasis on political-institutional over cultural differences (Pfau-Effinger 2005) and the omission of Southern (Ferrara 1996) as well as Eastern European patterns (Manning 2004), but also for its gender and family 'blindness' (Lewis 1992; Pfau-Effinger 1998; Sainsbury 1996), a particularly salient critique in the context of LTC and social support.[2] As a next step therefore, Anttonen and Sipilä (1996; see also Bettio and Plantenga 2004) developed a 'social care regime' typology consisting of four clusters based on an informal–formal dichotomy and considering both the dimensions of care for children and older people. Countries may then be clustered into (1) those that delegate care-giving responsibility almost entirely to the family (e.g. Italy, Greece, or Spain); (2) those that place great importance on informal care in general but treat childcare and eldercare differently by privatising childcare and providing some public eldercare (e.g. the United Kingdom, the Netherlands); (3) those giving priority to informal care strategies but offering substantial public compensation by collective arrangements, as is the case in Germany or Austria; and (4) the Nordic countries, which provide moderate to high levels of formal care to both older people and children.

In order to better understand drivers of and obstacles for social innovation and active ageing policies, to contextualise our findings and to do justice to the huge differences within and between EU Member States, it was necessary in our study to go one step further and elaborate on existing concepts of care regimes (Lamura 2007; Nies et al. 2013; see Table 9.1). Although such categorisations are undoubtedly subject to

Table 9.1 Care regimes as a context for social innovation and active ageing policies

	Demand for care	Provision of informal care	Provision of formal care	Acknowledgement of LTC as a social risk	Countries[a]
Standard-care mix	Medium-high	Medium	Medium	Early movers	**Germany, Austria,** France, United Kingdom
Universal-Nordic	Medium	Low	High	First movers	Denmark, **Finland, The Netherlands,** Sweden
Family-based	High	High	Low	Late movers	Spain, **Italy, Portugal,** Ireland, Greece
Central & Eastern European (CEE)	Low-medium	High	Low	Starters	**Hungary,** Poland, **Czechia,** Slovakia, **Romania, Bulgaria, Estonia,** Latvia, **Lithuania**

Source: [a]Further developed based on Lamura (2007), Nies et al. (2013)
Note: Ideal-type examples, countries addressed in the study in bold letters

ample discussion depending on definitions and conceptual approaches, some broad clusters of countries can be identified to distinguish differences in demand, supply, and policy frameworks. Using the variables 'demand for care', 'provision of informal care', 'provision of formal care', and 'acknowledgement of LTC as a social risk', four different care regimes can be identified:

(a) The Nordic countries (e.g. Finland) and the Netherlands which were first to acknowledge LTC as a social risk that calls for solidarity and universal coverage ('Universal-Nordic') could be called 'first movers' in the context of LTC.

(b) Countries (e.g. Austria, Germany) with a 'Standard care-mix' could still be called 'early movers' as they started to implement instruments to fund and develop LTC towards the end of the last century.
(c) The Mediterranean countries (e.g. Italy, Portugal) have been coined 'Family-based' as the (legal) responsibilities of families to provide care are most important, even if also in all other countries family care and subsidiarity principles remain the backbone of LTC provision.
(d) Finally, during the transition from communist regimes to market economies most CEE countries (e.g. Hungary, Romania) have undergone profound changes in social security systems. LTC was a non-priority for a long time but in more recent years different stakeholders have raised the need to set up priorities and clear regulations for the LTC sector, as the rapidly ageing population will increase demand for LTC services in the near future. Still, these countries could be called 'starters' in LTC.

It should be underlined that the notable similarities between the countries of each cluster do not preclude specific idiosyncrasies and differences, especially in terms of care demand, policy approaches, funding mechanisms or the levels of (de)familisation and commodification of care (Meagher and Szebehely 2013; Rodrigues et al. 2012; Rostgaard and Pfau-Effinger 2011; Simonazzi 2009). For instance, Italy's proportion of public expenditure on LTC is higher than in the United Kingdom and Austria, due to a widespread coverage of a flat-rate cash benefit in Italy, while the proportion of older people living in residential care is higher in Hungary than in Italy.

Further caveats for any comparison of LTC policies consist in population heterogeneity, political structure, and professional practice that are affected by the particular society in which they operate and reflect related historical, religious, and cultural forces. Yet the key challenges mentioned above are found to a lesser or greater degree in all EU countries and there is a clear need for change in LTC policy and practice in Europe. Our approach in this project was to confront these challenges as an opportunity to be innovative and as a catalyst for active ageing policies and social innovation. The next section is therefore dedicated to a brief definition of these concepts as they relate to LTC.

The Concepts of Social Innovation and Active Ageing in Relation to LTC

Social Innovation

The European Commission, in collaboration with the UK's Young Foundation, defines social innovation as new and progressive ideas, products, services, and processes that meet social needs by building new relationships or new forms of collaboration (European Commission 2011). Moulaert and colleagues (Moulaert et al. 2013) emphasise the role of new social relations in the definition, arguing that it is these new forms and channels of collaboration between diverse actors—from within the public, private, and civil society spheres—that lie at the heart of social innovation theory and practice. Also Howaldt et al. (2016, p. 1) stress the new 'constellations of stakeholders' and the 'new combination or new configuration of social practices in certain areas of action … with the goal of better satisfying or answering needs and problems than is possible on the basis of established practices'. In any case, there is growing interest in social innovation within the EU social policy discourse, as evidenced by the establishment of numerous international research networks, for example, the European Innovation Partnership on Active and Healthy Ageing (EIP AHA) and Social Innovation Europe as well as by funding streams allocating resources to address the need for social innovation, such as the Commission's Horizon 2020 framework funding programme.

In the context of LTC, however, social innovative practice needs to be further specified as proposed by Heinze and Naegele (2012). In identifying social innovation in the area of LTC, initiatives need to fulfil at least two of the following criteria:

- Orientation towards exceptional societal challenges or social issues;
- Suggestion of new solutions in the respective societal, cultural and economic context;
- Creation of new patterns of social practices to overcome shortcomings of traditional arrangements;

- Overcoming the traditional dichotomy between technological and social innovations;
- Promotion of the integration and/or collaboration of heterogeneous stakeholders that have hitherto not co-operated;
- Inclusion of reflective and multidisciplinary approaches towards the key goal of societal usefulness;
- Creation of structures and processes that are sustainable and realise new growth potential in terms of regular employment;
- Involvement of end-users as co-producers of services or products;

Finally, social innovation may take place on different scales. Initiatives may focus on a micro-scale development, such as changing relations between individuals and small groups, or they may be designed on a macro-scale (relations between groups) leading to systemic change (Garcia et al. 2010). This implies considering differences between countries or 'care regimes' not only in their approach to LTC, but also in the degree to which they support the circumstances necessary to nurture social innovation in LTC specifically.

Active Ageing

As with social innovation, at first glance the concept of '*Active Ageing*' does not seem to go hand in hand with LTC because it is traditionally associated with labour force participation and/or volunteering. The general WHO definition goes beyond this typical understanding. Rather active ageing is:

> ... the process of optimizing opportunities for health, participation and security in order to enhance quality of life as people age. It applies to both individuals and population groups. Active ageing allows people to realize their potential for physical, social, and mental well-being throughout the life course and to participate in society, while providing them with adequate protection, security and care when they need. The word 'active' refers to continuing participation in social, economic, cultural, spiritual and civic affairs, not just the ability to be physically active or to participate in the labour force. Older people who retire from work, ill or live with disabilities

> can remain active contributors to their families, peers, communities and nations. Active ageing aims to extend healthy life expectancy and quality of life for all people as they age (WHO 2016)

People in need of LTC are by definition, and particularly in the case of severe physical, mental or cognitive impairments, challenging the active ageing paradigm, even if empowerment and enabling strategies that help older people (and their relatives) and make the most out of their respective situation are enshrined in this framework. Active ageing approaches must therefore include strategies to extend healthy life expectancy and quality of life for *all people* as they age including those who are frail, disabled, and in need of care (Walker and Foster 2013; Foster and Walker 2015). Promoters of 'Active (and Healthy) Ageing' have conspicuously taken on board the issues of 'cure and care' (see e.g. the European Year 2012, but also the European Innovation Partnership on Active and Healthy Ageing or the Seventh German Federal Report on Ageing, BMFSFJ 2017). Still, it should be avoided that active ageing turns into a prescriptive norm that discriminates against those who, for whatever reason, are not able to age in an active and healthy manner (Angus and Reeve 2006; Ruppe 2011).

Active ageing in the context of LTC is, perhaps more than in other areas, dependent on social investment strategies across a range of policy fields. For instance, there are many possibilities for organisational improvements in terms of coordination with health systems, for promoting 'carer-friendly' employment policies and for designing user-friendly information and communication technology (ICT) applications to enable social participation and solidarity for older people in need of LTC. Apart from obligations to take advantage of these opportunities and remain active, these strategies have to be based on rights to social protection and other forms of social welfare, thus combining top-down and bottom-up initiatives that enable people to develop their own forms of activity.

Based on the clarification of these two central concepts—social innovation and active ageing—it was possible to isolate opportunities for and ongoing practices of social innovation and active ageing within the LTC systems of selected EU Member States. The research team identified more than 60 initiatives in the selected countries (AT, BG, CZ, DE, EE, FI, HU, IT, NL, RO, PT), 18 out of which were analysed as case studies based on qualitative research methods (Schulmann and Leichsenring 2015). Table 9.2 provides an overview of some key innovative features and potentials that will be further analysed in the subsequent sections.

Potential for Active Ageing through Social Innovation in LTC and Social Support

The following sections will outline our key findings; first, with respect to drivers and barriers for social innovation in LTC and, second, in relation to a limited number of innovative and realistically implementable scenarios of LTC and social support in EU Member States (Leichsenring et al. 2016b). Future scenarios consider the experiences of existing good practice by 'establishing and expanding LTC as a system', 'shifting LTC to community-based care', 'facilitating integration and coordination of LTC systems' as well as 'creating and improving employment'.

Drivers and Barriers of Social Innovation in LTC

The relatively small number of social innovation initiatives in LTC may be explained by select key factors identified through the analysis of expert interviews and focus groups with relevant stakeholders involved in the development, implementation, and diffusion of often small-scale initiatives and projects. Table 9.3 presents these key factors as well as related drivers and barriers. Based on a literature review and the case-by-case analysis of innovative initiatives, it could be shown that a range of dynamic factors at the macro-, meso- and micro-levels influence innovative practices. Not surprisingly, sustainable initiatives have public support, both in terms of financing and regulatory frameworks, they encourage the creation of new institutional partnerships, build networks, and reach across professional, disciplinary and sector boundaries. Working towards an integrated service model in which users' care needs and resources are managed in a coordinated way is particularly relevant for social innovations in LTC.

Table 9.2 Case studies of social innovation and active ageing in LTC, by care regime

Care regime	Social innovation case studies	Main innovative feature(s)	Status
Standard-Care-mix	Bielefelder Model, DE	Moving from single care services to inclusive care in the neighbourhood with the idea of a 'caring community'	Regional practice, ongoing
	Care Support Centres in Mönchengladbach, DE	New practice model; expansion of existing structures; new partnerships between diverse stakeholders	Regional practice, ongoing (limited areas)
	Village Service, AT	New practice model; new partnerships between diverse stakeholders; new employment opportunities; grassroots	Regional practice, ongoing (limited areas)
Universal-Nordic	Buurtzorg (Care in the neighbourhood), NL	New professional profile and employment opportunity; new partnerships between formal and informal stakeholders; grassroots	Mainstreamed practice, ongoing; transferred to other countries
Family-based	Ageing Geo-Information System, PT	New partnerships between diverse stakeholders; Smart use of ICT for care planning	National practice, ongoing
	Active Ageing with Dementia, PT	Involves end-users in development/ implementation; creates new roles and partnerships	Regional practice, ongoing (initially national pilot)
	Care for Carers, PT	New partnerships between diverse stakeholders; user involvement in development/ implementation	Regional practice, ongoing (limited areas)
	Family Nurse Program, IT	New professional profile; new practice model	Regional practice, ongoing
	Recognition of informal skills, Piedmont region, IT	New practice model; new employment opportunity; new partnerships between diverse stakeholders	Regional practice, ongoing
	UP-TECH project, supporting caregivers of Alzheimer's disease patients, IT	Promoting integration and collaboration of heterogeneous stakeholders that have hitherto not co-operated; incubator with multidisciplinary approaches	Regional practice, ongoing

(continued)

Table 9.2 (continued)

Care regime	Social innovation case studies	Main innovative feature(s)	Status
CEE	AREÍON Emergency care, CZ	New service/product; new practice model	National practice, ongoing
	CarEIn, Emotional intelligence care in health and social assistance, BG	New practice model; new employment opportunity	National pilot, terminated
	Carer+, Developing Digital Competencies of Care Workers to Improve the Quality of Life of Older People, RO, LV	New practice model; smart use of ICT	International practice, ongoing
	Elderly-friendly Housing, HU	New practice model	Regional practice (limited areas
	Home Care and Assistive Services for an Independent and Dignified Life, BG	New practice model; new professional profile; New partnerships between diverse stakeholders	National pilot, ongoing (limited areas)
	Integrated Help-at-Home Development Program, LT	New partnerships between diverse stakeholders; new social networks	National pilot (limited areas), terminated
	VIRTU-Virtual Elderly Care Services on the Baltic Islands, EE, FI	New practice model; smart use of ICT	Regional practice, ongoing (limited areas)
	WebNurse, online training for informal carers, HU	Smart use of ICT	National practice, ongoing

Source: Schulmann and Leichsenring (2015)

Table 9.3 The key factors involved in social innovation in LTC, its drivers and barriers

Key factors	Drivers	Barriers
Coordination/ integration	Uses integrated care model	Structural fragmentation of LTC system; Lack of coordination between partners
Design	Evaluation is incorporated into design; Efficient use of ICT; Universal access; Expert input and feedback; Successful dissemination; Quality management system is incorporated; Rigorous evidence base	Lack of underlying incentive structure; Difficulties recruiting participants; Difficulties evaluating the initiative; Design is ill-suited to meet needs
Framework/ structural conditions	Legislative foundation/ recognition of services provided; Draws on existing resources (e.g. human resources, existing built infrastructure); Autonomy of affiliate organisations	Unfavourable framework/ structural conditions; Lack of harmonised data; Ill-defined identity of the initiative (e.g. legal status); Disinterest on the part of policymakers
Funding	Affordability for the end user; Raising private funds; Public sector co-financing; EU-level funding	User payment required; Insufficient funding (public or private); No public funding; High implementation costs; Difficulties securing transition from EU- to public funding
Leadership	Institutional leadership (often by an organisation in the third sector)	Lack of leadership
Local/ community focus	Adapted to meet local needs and contexts; Strong sense of community ownership; Broad community involvement	'One size fits all' approaches
LTC specificity	Incorporates a community care model; Case management component; Incorporates individualised, user-centred care plan	No acknowledgement of LTC as a public priority

(*continued*)

Table 9.3 (continued)

Key factors	Drivers	Barriers
Network	Well-established/active stakeholder network; Public–private partnership; Contributions of volunteers; Formalised institutional partnerships; Multi-actor/multi-sector cooperation	
Sustainability	Successful transition from pilot programme; Integration of services into publicly provided services	Short duration leading to lack of continuity and sustainability; Lack of dissemination/awareness-raising
Target group	Well-defined target group	Restricted coverage; Resistance to participation; Lack of computer literacy
User involvement	User-led components; User input and feedback during development and implementation	
Workforce	Multidisciplinary project team; Built-in element to ensure workforce sustainability	Unskilled/ill-supported informal care workforce; Insufficient (human) resources

Source: Authors' compilation

Areas with the Greatest Potential and Need for Social Innovation in LTC

Establishing and Expanding LTC as a System

Even though the expansion of services as such cannot be seen as social innovation, many initiatives analysed in this study strive to combine new approaches, new relationships between stakeholders or technological applications to reduce gaps in access for marginalised target groups as well as gaps in professionalisation and communication. The expansion of LTC services represents a priority in all selected countries, particularly in rural areas, and even in countries with so-called well-developed LTC infrastructures, but the necessary preconditions do not exist everywhere (Széman et al. 2016). Although the expansion of LTC infrastructure could represent an area for job creation and the modernisation of outdated institutions, there is a lack of communication and funding in CEE countries. Although ICT is expanding in Care-mix countries, the potential for broad implementation still has not been utilised and needs to be better combined with social innovation that, in many cases, would just consist in the involvement of users and other stakeholders. For instance, younger generations and the wider community were successfully integrated in service expansion in Finnish-Estonian ('VIRTU') and German ('Care Support Centres in Mönchengladbach') initiatives. Other opportunities consist in the integration of migrants not only as carers but also as clients of LTC services.

The potential for improvement through service expansion also entails a better definition of target groups in need of LTC to avoid undersupply, inappropriate, or in some cases even oversupply. Users should get needs-based access independently from the origin of their disability because different regulations for specific age groups, diseases, or disabilities are causing a lack of coverage and appropriate support. Acknowledging informal carers as a specific target group for support measures may contribute to a better work–life-care balance for carers of working age, and older carers (e.g. spouses) could be protected from the kind of strain that results in their needing LTC services themselves. A better balance of service provision between metropolitan and rural areas, for example, by means of appropriate financial incentives for municipalities and professional interventions at the local level, could (re-)strengthen social ties in rural areas, neighbourhoods, and communities, as further described in the following section.

Shifting LTC Systems Towards Community-Based Care

In general, the aim of community care is to enable the user to live at home as independently as possible, for as long as possible, and to delay admission to residential care, that is, to a nursing home ('ageing in place'). In the context of this study, a distinction is made between community care (for definition, see Kassner 2011) and *community-based care*. Community-based care means that the locality—with disparate stakeholders working together—serves as the nexus of support and care provision, and as a place of confluence for formal and informal support, including volunteer services.

The examples of good practice identified in the different care regimes show considerable potential for reshaping LTC services to better meet the needs of older adults and the people closest to them because community-based care offers a framework for care provision that shifts the focus from an institutional setting to the user's home in a way that is both cost-effective and beneficial for users. In addition, community-based care forms the basis for better integration of the various stakeholders and improved coordination among care providers across the informal–formal care divide. These features of community-based care are achieved to a remarkable extent by the 'Integrated Help-at-Home Programme' in Lithuania and 'Buurtzorg' in the Netherlands (Table 9.2).

The research also showed that a distinction emerges between those countries that have already started to realise community-based care innovations (Care-mix, Universal) and those in which there remain considerable opportunities to do so (CEE, Family-based). Yet even among the former, community-based care models are not mainstream practice and tend to be isolated to specific localities and regions. Transferability of best practices in this area is hampered by existing framework conditions that require concerted reforms at the national, regional and local levels. It is therefore recommended to endow local governments with the authority to plan and coordinate care services, with at least partial long-term funding support from national sources. In addition, it seems necessary to incentivise civil society initiatives with financial support that extends beyond the traditional duration of pilot programmes. This entails systematic evaluation and a strategy for scaling up into innovation programmes receiving public funding. At the institutional level, it is necessary to develop community networks and create time and space for negotiation with all stakeholders involved. Establishing service planning based on needs assessments that involve users should also be prioritised, as should the recruitment of personnel and volunteers from within the community.

Furthermore, decentralising care planning and organisation has the advantage that the people with decision-making power tend to have a more nuanced understanding of patterns of care need and institutional capabilities in their local context than do external actors. While the embeddedness in local infrastructure is a key to community-based care, one of its crucial preconditions is appropriate regulatory and financial support from national institutions. Social investment is therefore needed to tap into the potential to create employment and to embark also on ICT-supported care delivery that can be accessed and accepted by all members of the community.

Facilitating the Integration and Coordination of LTC Systems

The need for more coordinated or even integrated policies that facilitate the '… provision of health care, social services and related supports at the right time and place' (Kodner and Spreeuwenberg 2002, p. 3) has been a recurring issue in developing LTC systems across Europe over the past few decades (Leichsenring 2004; Ham and Smith 2010; Kodner 2009; WHO 2015; Van der Heide et al. 2015). The current fragmentation of services has been identified as a problem by users and informal carers in this study as well. Also providers of facilities and services have repeatedly criticised fragmented funding structures and regulatory frameworks in interviews and focus group discussions, while experts highlighted the expected benefits of integrated care in terms of the 'triple aim' of current health care reforms—to achieve better quality and better health at reduced costs. Apart from the divide of health and social care, and peculiar partitions in governance and delivery structures, it is in particular the divide between formal and informal care that has been identified as a huge societal challenge (Colombo et al. 2011). This situation and the wide variety of approaches to mitigate evident shortcomings in LTC contain high potential for social innovation and improvements concerning better access and appropriate service provision, in particular for people with multimorbidity and LTC needs at higher ages.

Focusing on integration and coordination, the results of 14 out of 18 case studies of innovative practices were analysed to identify the key success factors of social innovation that have shown evidence to realise better coordination or even integrated service delivery (Leichsenring et al. 2016a).

The findings revealed that extending the scope of coordination and networking and the related involvement of (new) stakeholders are crucial factors to realise social innovation. Social innovation with a view to coordination and integration is increasingly relying on broader societal solutions that include, for instance, mobile health (mHealth) and other ICT applications, housing issues, employment, and general issues related to ageing as a cross-sectoral issue of local policies. At first sight, this seems to contradict integration and coordination as decision-making structures and processes are likely to increase in complexity with the rising number of players.

However, results from case studies show that expanding the networks and including hitherto 'outsiders' (e.g. volunteers) in the delivery of social support and LTC offers opportunities for new types of solutions. Various innovations indicated a tendency towards activating local resources and networks to strengthen and renovate LTC provision, such as the 'Bielefelder model' and the 'Care Support Centres', which clearly hint at the important role of neighbourhoods in Germany. The 'Buurtzorg' model in the Netherlands heralds this trend by its very name, conceptually building on the professional activation of clients' resources in their local environment by skilled community nurses. Similar approaches could also be found in the Italian action research project 'UP-TECH', where case management and smart ICT were used as instruments to raise local awareness and to involve local stakeholders, and in Portugal, where local private enterprises were involved in funding and promoting the initiative 'Active Ageing with Dementia'. All these initiatives demonstrate that participation of and consultation with relevant stakeholders in the local context are crucial from the very beginning of any initiative in order to increase acceptance, mutual trust, and sustainability. Local authorities must therefore develop cooperative structures to foster, strengthen, or at least to not hamper bottom-up initiatives, for example, by organising regularly 'Round Tables' with pertinent stakeholders. In rural areas, this might include the creation of alliances between small municipalities. For instance, 'Care Support Centres' (Germany), a case management and care consultancy initiative, can only be successful in a sufficiently developed environment of care providers, with appropriate professional resources to arrange 'care packages' and detect shortcomings that can then be addressed by the involved stakeholders. The latter function could also be applied in CEE countries, for example Romania, but only if sufficient resources were in place to follow up and expand services.

Professional management at the numerous interfaces between health and social care, but also between formal and informal care, is in any case a necessary precondition for addressing new societal challenges in LTC. This is why new job profiles such as case managers, network coordinators, social animators, and mediators have become key in successful social innovation initiatives. Good examples to illustrate innovative responses to social challenges were found in Family-based care regimes. For instance, the Piedmont region in Italy started integrating live-in migrant carers into the formal labour force by acknowledging informal skills acquired through practical caregiving experiences. And the Portuguese initiatives 'Active Ageing with Dementia' and 'Care for Carers' are both initiatives born out of the recognition of specific needs that were addressed by inter-municipal, multidisciplinary and public partnerships, partly also involving private enterprises.

A major issue in coordinating LTC and social support is striking a balance between national framework legislation (funding, standard setting, controlling) and the local settings in which services, facilities, and care relations are being organised and delivered. Even countries with a somewhat longer tradition of LTC legislation (Germany, Austria, the Netherlands, Finland) still struggle with the division of responsibilities between health and social care and between different levels of governance. This same division is also evident in CEE countries where the divide between health and social care seems to be even more articulated, and in Family-based care regimes. National governments need to embrace the concept of 'social investment', in particular in those countries where social support and LTC policies have not yet been fully implemented. This needs to be underpinned by a profound inter-sectoral dialogue to shape appropriate national framework conditions and by tangible 'Social Innovation Funds for LTC' or similar reform programmes with the aim of promoting change and improvement. In this connection, coordination between national authorities and EU-funding opportunities needs to be enhanced as initiatives by local partnerships based on funding by EU programmes often do not find subsequent financial backing at the national level beyond the pilot phase.

Creating and Improving Employment in LTC

Employment represents a crucial component of LTC provision in ageing societies as the growing demand for care calls for an increased availability of qualified care professionals in all care regimes. While informal carers will certainly remain the backbone of LTC across all countries, professionalisation and employment in LTC are largely triggered by public funding. However, both Care-mix regimes, where employment in LTC is generated by dedicated funds to (co-)finance residential facilities and community care, and the Nordic countries with universal coverage, have shown that in particular the extension of services in-kind contributes to an extension of employment with a clearly visible social return on investment. This should be seen as a potential also for CEE countries and the Family-based care regimes.

The general trend towards a 'commodification of LTC', however, calls for a more ample societal debate. The term not only describes the tendency to a 'professionalisation of care' (Rostgaard and Pfau-Effinger 2011) and thus the extension of paid care work but also a more market-driven care provision and the emergence of (new) types of providers. Furthermore, the structural distinction between health and social care responsibilities has resulted in a wide range of professional profiles and specific skills to carry out legally delineated tasks. As a result, many countries are confronted with increasing fragmentation and an increasing shortage of trained staff.

The amalgam of a lack of services and the increasing participation of women in the labour market on the one hand and low wages and unemployment in the CEE countries on the other has triggered another phenomenon that needs to be addressed at the European level. Migrant carers have not only contributed to a new mix of professionals in the formal care system, but in particular to new types of arrangements in a grey area between formal and informal care work, including personal care provided by live-in, mainly female migrant carers. This type of social innovation has spread at an impressive pace for a number of reasons, including the existence in certain countries of cash-for care benefits used to privately hire these carers (Beltrametti 2008). The migration of care workers, be it in the informal or formal care sector, while providing a solution to staff shortages in the destination countries, exacerbates the shortage of (formal and informal) carers in the sending countries, largely represented by CEE and non-EU countries (Williams 2012).

Other cross-cutting issues with clear relevance at the micro- and meso-levels which emerged in this study refer, in the first place, to the need to develop professional competencies to properly support older people in rural areas. One emerging proposal in this respect envisages the re-introduction of the so-called district nurses, who in the past ensured appropriate care for older residents in cooperation with physicians and municipalities. Another much needed intervention is training LTC staff in the use of new technologies. The positive impact of technology-based innovations is hindered if the digital literacy of the target user—and of care staff—is not appropriately evaluated (Rossi Mori and Dandi 2012). This is a frequently occurring phenomenon, with the result that a tool that has the potential to be a driver of innovation can actually become a barrier, depending on the context. Finally, LTC professionals often have insufficient knowledge about the care provided by other professionals within the LTC system, adhering to a silo-like approach that results in a lack of cooperation between professional groups. The combination of narrow specialisation without appropriate coordination, especially in times of increasing differentiation in care options, represents an area in which social innovation is urgently needed.

The impact of the trends and phenomena described above on individual countries' LTC employment varies widely across Europe as educational profiles, qualifications, and training contents required of and available to LTC workers differ often even within countries (Geerts 2011). While in some care regimes they are regulated at national (macro) level, in others they are governed by regional (meso) authorities.

Against this backdrop, this study has identified a range of cross-cutting 'enablers' of social innovation which might contribute to create and improve employment opportunities in the field of social support and LTC across the different care regimes (Casanova et al. 2016). In the

Family-based care regimes, Spain has taken the most explicit approach to make LTC an area of job creation and greater service orientation by the introduction of a 'System for Autonomy and Care for Dependency' (SAAD). However, also due to the economic and financial crisis, efforts to extend the quantity and quality of care via staff development initiatives remain scattered and need to be streamlined to enhance support for family carers and training of low-skilled migrant carers, for example, by using web-based E-learning platforms. In CEE countries, the public sector remains hesitant to fund LTC services. As a result, non-governmental organisations (NGOs) have started to act as a bridge between public authorities and people in need of LTC. Their initiatives, ranging from training programmes for LTC professionals, managers, and volunteers to needs assessment and E-care projects, are often based on EU funding but would be in need of mainstream funding. In the Care-mix countries, innovations to strengthen the role of rehabilitation and case management should be further rolled out (e.g. 'Care Support Centres' in Germany). In this context, the implementation of more appropriate job profiles in LTC (e.g. similar to the 'Professional Social Carers' in Austria) would contribute to developing a distinct identity for LTC workers with appropriate skills for the complex tasks to be fulfilled. While the proportion of health and social care professionals in overall employment remains the highest in Europe in Universal care regimes, these frontrunners of employment strategies in LTC have also started to 're-discover' family care. In this context, it will be necessary to improve working conditions in the tradition of supportive welfare services, to develop company measures for the reconciliation of work and care and to strengthen partnerships.

Social innovations in the LTC employment sector also call for recommendations for EU-level policy makers. Aside from one of the most important and often repeated recommendations—to improve working conditions in the (formal) LTC sector—it would be necessary to develop large-scale programmes to train and re-train staff on how to best tackle chronic conditions via prevention, rehabilitation, and home care (Council of the European Union 2014). This is crucial in Family-based and CEE countries, where LTC is often seen as low-skilled work. Also in other care regimes the introduction of holistic training modules in geriatrics and LTC for all health and social care professionals, including GPs, is recommended, for example via Continuing Medical Education (CME) modules. Additionally, proactive strategies concerning cross-border movement of the LTC workforce and (migrant) carers are needed with respect to both the supply of care labour and the patterns of care demand. Such a strategy would be to extend the European professional card according to the EU Directive 2013/55 to all LTC professions.

In sum, promoting employment in LTC means in all care regimes, on the one hand, setting up quantitatively appropriate infrastructures to address care needs and, on the other hand, investing in human resources to pursue quality of care and thus to ensure a dignified life for older people and their informal carers. All these factors are strongly related to the promotion of social innovation in the LTC sector and to stimulating active ageing through social investment. Extending employment in LTC will certainly imply public expenditure for both professional staff and informal (family) carers, including all types of training and monitoring, but also to improve the image of LTC, to attract workforce and to support the intrinsic motivation of LTC provider organisations.

Conclusion

This study collected and analysed examples of social innovation practice in LTC from a wide range of countries. The findings raise two main points of interest—the heterogeneity in the type of innovations and the degree to which experiences and know-how across care regimes can be or are likely to be transferred.

There is a fundamental divide between the countries under review in terms of the social innovation practices identified. The Netherlands as the representative of the Universal-Nordic care regime, and Germany and Austria in the Standard-Care-mix regime are ahead of the others in terms of setting the 'state of the art' in social innovation practices. They are also at the forefront, together with the United Kingdom among others, of research into the theoretical conceptualisations of social innovation. It is evident from the analysis of the innovation case studies that what is considered innovative in CEE countries—the emergency alarm system 'AREÍON' from Czechia, or the 'WebNurse' initiative from Hungary—is not considered innovative in countries like the Netherlands or Austria. Yet in the CEE countries, where the level of familisation is high (e.g. in Hungary, the constitution explicitly states that care for older people is the legal responsibility of family members), where public expenditure on LTC is low in relative terms, and where in the not-so-distant past the private and third sectors were intentionally squeezed out by the state, a web-based training tool for informal carers is indeed innovative. Variation across countries and care regimes with regard to innovative LTC practices/interventions is linked to the level of development of countries' LTC systems, and thus countries in an 'LTC innovation typology' would fall along similar lines as the care regime typology presented in Table 9.1.

Given the 'starter' status of CEE countries and 'late mover' status of the countries in the Family-based care regimes—in terms of the recognition of LTC as a social risk and the level of development of their LTC systems—and in parallel, the lag witnessed in these countries in terms of the level of innovation relative to the 'first movers' and 'early movers', the question of what happens next looms large. In order for innovations to be developed, successfully implemented, rigorously evaluated and ultimately scaled up, considerable public investment is required. Investment in terms of funding but also, as a first step, in the creation of regulatory frameworks that enable bottom-up innovations to thrive. It remains to be seen whether the countries of the CEE care regime in particular will make this kind of social investment a priority. Considering that most countries are already struggling to meet the demand for LTC and are in some cases cutting public expenditure, it seems unlikely that the kind of investment necessary from the public sector will be forthcoming in the near future. The irony is that the push for 'social innovation' originates in the very need for alternative, cost-saving solutions to meet societal needs in economically precarious times.

A second issue raised by this study is to what extent countries, and again particularly countries in the CEE care regime, can learn from the experiences of countries with more well-established LTC systems, thereby avoiding similar pitfalls. In the countries of the Universal-Nordic and Standard-Care-mix regimes, the road to the development of LTC systems began with a patchwork of services, some previously provided under the auspices of health care and some within the domain of social care, until the need for a distinct LTC system with distinct services was acknowledged and accomplished to varying degrees of success through increased integration and coordination of care planning and provision. It would be gratifying if countries just beginning to develop their LTC systems could avoid the fragmentation experienced by frontrunner or 'first mover' countries, and by implementing innovative care models, 'disrupt' the pathway to arrive at a well-functioning LTC system that meets the needs of the population.

One way to make this happen is to place greater emphasis on improving the evidence base for innovations and good practices that have demonstrated promising results. Rather than reiterating the need for new and innovative approaches and interventions, greater emphasis should be placed on replicating promising innovations that have been developed and implemented in certain countries, and seeing how these might be transferred to other contexts and what kind of modifications this might require.

A last key point relates to the term 'social innovation' itself, and to its visibility and comprehensibility outside policymaking circles. The interviews and focus groups conducted with care professionals and family members of older people in the course of this study demonstrate a disconnect between policy and practice. Social innovation is not a term that most practitioners are familiar with; instead, they speak in terms of problems they encounter in their day-to-day working lives and suggest improvements based on the resources at their disposal. If investing in social innovation is to be a successful policy care providers need to be brought into the conversation, and what is meant by 'social innovation' needs to be clarified and refined.

To further stimulate social innovation and active ageing in LTC, EU institutions have an important role to play in supporting Member States. Apart from promoting a common understanding of national services and guidelines for LTC and the support of isolated projects, it would be necessary to follow-up and scale-up positive experiences in collaboration with NGOs and local authorities, in particular in CEE countries. Transferability of good practice between countries will remain a major challenge if, as innovative initiatives revealed in some cases, it has been difficult even to exchange results between regions in the same country. Further research on outcomes in terms of equity and person-centredness and accompanying initiatives showing their results may support such processes. Another area for future research will be to address the 'dualisation' of European integration, with 'poorer' Member States keeping LTC in 'richer' Member States viable.

Acknowledgements The authors would like to thank and acknowledge the contributions to this chapter by Georgia Casanova and Giovanni Lamura, National Institute of Health and Science on Ageing (INRCA), Ancona, Italy; Mihaela Ghenta, Speranta Pirciog, Aniela Matei and Luise Mladen, National Research Institute for Labour and Social Protection (INCSMPS), Bucharest, Romania; Niku Määttänen and Tarmo Valkonen, The Research Institute of the Finnish Economy (ETLA), Helsinki, Finland; Gerli Paat-Ahi and Vootele Veldre, Praxis Centre for Policy Studies Foundation, Tallinn, Estonia; Gerhard Naegele and Sandra Schulze, Technical University Dortmund, Institute for Gerontology (TUD IFG), Dortmund, Germany; Anett Mária Tróbert and Zsuzsa Széman, Institute for Sociology, Research Centre for Social Sciences, Hungarian Academy of Sciences (IS RCSS HAS), Budapest, Hungary; Rita Gouveia and Karin Wall, Institute of Social Sciences, University of Lisbon (ICS), Lisbon, Portugal.

Notes

1. For the purpose of our study LTC has been defined by considering a number of criteria to focus on the continuity of social and health care with the aim to overcome barriers at the interfaces between social and health care, and between formal and informal care (Schulmann and Leichsenring 2014; Leichsenring et al. 2013).
2. For a critical overview of the 'welfare regime' approach, see Hoff (2006).

References

Anderson, A. (2012). Europe's Care Regimes and the Role of Migrant Care Worker Within Them. *Journal of Population Ageing, 5*(2), 135–146.

Angus, J., & Reeve, P. (2006). Ageism: A Threat to 'Aging Well' in the 21st Century. *Journal of Applied Gerontology, 25*(2), 137–152.

Anttonen, A., & Sipilä, J. (1996). European Social Care Services: Is It Possible to Identify Models. *Journal of European Social Policy, 6*, 87–100.
Beltrametti, L. (2008). Il Finanziamento). In C. Gori (Ed.), *Le riforme regionali per i non autosufficienti* (pp. 47–73). Rome: Carocci.
Bettio, F., & Plantenga, J. (2004). Comparing Care Regimes in Europe. *Feminist Economics, 10*, 85–113.
BMFSFJ (Bundesministerium für Familie, Senioren, Frauen und Jugend; Federal Ministry for Family, Seniors, Women and Youth). (2017). Altenbericht der Bundesregierung. Sorge und Mitverantwortung in der Kommune – Aufbau und Sicherung zukunftsfähiger Gemeinschaften. *(7. German Federal Report on Ageing- Caring and shared responsibility in the Community.*
Casanova, G., Schulze, S., Mladen, L., Naegele, G., Leichsenring, K., & Lamuara, G. (2016). Social Innovations Facilitating the Creation and Improvement of Employment Opportunities in Social Support and LTC. Policy Paper #4. In K. Leichsenring, K. Schulmann, G. Casanova, M. Ghența, G. Lamura, A. Matei, … M. A. Tróbert (Eds.), *Potentials for Social Innovation in Social Support and Long-Term Care* (pp. 42–56). A Series of Policy Papers. Vienna: European Centre. MoPAct Report, #8.4.
Colombo, F., & Mercier, J. (2011). Help Wanted! Balancing for Protection and Financial Sustainability in Long-Term Care. *Eurohealth, 17*(2–3), 3–6.
Colombo, F., Nozal, A.-L., Mercier, J., & Tjadens, F. (2011). *Help Wanted? Providing and Paying for Long-Term Care, OECD Health Policy Studies.* Paris: OECD Publishing.
Council of the European Union. (2014). *Adequate Social Protection for Long-Term Care Needs in an Ageing Society.* European Union. Retrieved from https://ec.europa.eu/health//sites/health/files/ageing/docs/ev_20140618_co04_en.pdf
Deusdad, B. A., Pace, C., & Anttonen, A. (2016). Facing the Challenges in the Development of Long-Term Care for Older People in Europe in the Context of an Economic Crisis. *Journal of Social Service Research, 42*(2), 144–150.
ENNHRI -European Network of National Human Rights Institutions. (2017). *Long-Term Care in Europe.* ENNHRI. Retrieved from http://ennhri.org/Long-term-Care-in-Europe
Esping-Andersen, G. (1990). *The Three Worlds of Welfare Capitalism.* Oxford: Polity Press.
European Commission. (2011). *Empowering People, Driving Change: Social Innovation in the European Union.* Luxembourg: European Commission.
Eurostat. (2015). *People in the EU – Population Projections.* Retrieved March 20, 2018, from Eurostat. http://ec.europa.eu/eurostat/statistics-explained/index.php/People_in_the_EU_%E2%80%93_population_projections#Europop2013_.E2.80.94_population_projections

Ferrara, M. (1996). The 'Southern' Model of Welfare in Social Europe. *Journal of European Social Policy, 55*(1), 17–36.

Foster, L., & Walker, A. (2015). Active and Successful Aging: A European Policy Perspective. *The Gerontologist, 55*(1), 83–90.

Garcia, M., Eizaguirre, S., & Pradel, M. (2010). Theorising Governance in Social Innovation Dynamics and Strategies. In F. Moulaert, D. MacCallum, A. Mehmood, & A. Hamdouch (Eds.), *Social Innovation: Collective Action, Social Learning and Transdisciplinary Research* (pp. 31–40). KARTASIS Final Report.

Geerts, J. (2011). *The Long-Term Care Workforce: Description and Perspectives.* ENEPRI. Retrieved from http://www.ancien-longtermcare.eu/sites/default/files/ENEPRIRR93_ANCIENWP3_0.pdf

Greve, B. (2016). *Long Term Care for Elderly in Europe.* Oxon: Routledge.

Ham, C., & Smith, J. (2010). *Removing the Policy Barriers to Integrated Care in England.* London: The Nuffield Trust.

Heinze, R., & Naegele, G. (2012). Social Innovations in Ageing Societies. In H.-W. Franz, J. Hochgerner, & J. Howaldt (Eds.), *Challenge Social Innovation* (pp. 153–167). Heidelberg: Springer.

Hoff, A. (2006). *Lone Mothers Between the Welfare State and Informal Support.* Lewiston: The Edwin Mellen Press.

Hoff, A., Reichert, H., Hamblin, K., Perek-Bialas, K., & Principi, A. (2014). Informal and Formal Reconciliation Strategies of Older Peoples' Working Carers: The European carers@work Project. *Vulnerable Groups & Inclusion,* 5, 1. https://doi.org/10.3402/vgi.v5.24264.

Howaldt, J., Kaletka, C., & Schröder, A. (2016). *Mapping the World of Social Innovation. Key Results of a Comparative Analysis of 1.005 Social Innovation Initiatives at a Glance.* Retrieved from http://www.sfs.tu-dortmund.de/cms/de/Aktuelles/170104_global_mapping_si_drive_innovationen/Mapping_the_World_of_Social_Innovation.pdf

Kassner, E. (2011). *Home and Community-Based Long-Term Services and Supports for Older People.* Washington, DC: AARP Public Policy Institute.

Kodner, D. (2009). Integrated Service Models: An Exploration of North American Models and Lessons. In J. Glasby & H. Dickinson (Eds.), *International Perspectives on Health and Social Care: Partnership Working in Action* (pp. 68–80). Oxford: Wiley-Blackwell.

Kodner, D. L., & Spreeuwenberg, C. (2002). Integrated Care: Meaning, Logic, Applications, and Implications – Discussion Paper. *Journal of Integrated Care, 2*, e12.

Lamura, G. (2007). *Dimensions of Future Social Service Provision in the Ageing Societies of Europe*. VI. European Congress of Gerontology and Geriatrics. St. Petersburg, Russia.

Leichsenring, K. (2004). Providing Integrated Health and Social Care for Older Persons – A European Overview. In K. Leichsenring & A. M. Alaszewski (Eds.), *Providing Integrated Health and Social Care for Older Persons – A European Overview of Issues at Stake* (pp. 9–52). Aldershot: Ashgate.

Leichsenring, K., Billings, J., & Nies, H. (Eds.). (2013). *Long-Term Care in Europe: Improving Policy and Practice*. London: Palgrave Macmillan.

Leichsenring, K., Casanova, G., & Schulze, S. (2016a). Facilitating Integration and Coordination in LTC. Policy Paper #2. In K. Leichsenring, K. Schulmann, G. Casanova, M. Ghenţa, G. Lamura, A. Matei, … M. A. Tróbert (Eds.), *Potentials for Social Innovation in Social Support and Long-Term Care* (pp. 16–27). A Series of Policy Papers. Vienna: European Centre. MoPAct Report, #8.4.

Leichsenring, K., Schulmann, K., Casanova, G., Ghenţa, M., Lamura, G., Matei, A., … Tróbert, M. A. (2016b). *Potentials for Social Innovation in Social Support and Long-Term Care*. A Series of Policy Papers. 56.

Lewis, J. (1992). Gender and the Development of Welfare Regimes. *Journal of European Social Policy, 2*, 159–173.

Lutz, H., & Palenga-Mollenbeck, E. (2010). Care Work Migration in Germany: Semi-Compliance and Complicity. *Social Policy and Society, 9*(3), 419–430.

Manning, N. (2004). Diversity and Change in Pre-accession Central and Eastern Europe Since 1989. *Journal of European Social Policy, 14*(3), 211–232.

Meagher, G., & Szebehely, M. (2013). *Marketisation in Nordic Eldercare. A Research Report on Legislation, Oversight, Extent and Consequences*. Stockholm: Stockholm University (Normacare).

Moulaert, F., MacCullum, D., & Hillier, J. (2013). Social Innovation: Intuition, Precept, Concept and Practice. In F. Moulaert, D. MacCullum, A. Mehmood, & A. Hamdouch (Eds.), *The International Handbook on Social Innovation. Collective Action, Social Learning and Transdisciplinary Research* (pp. 13–24). Cheltenham: Edward Elgar.

Nies, H., Leichsenring, K., & Mak, S. (2013). The Emerging Identity of Long-Term Care Systems in Europe. In K. Leichsenring, J. Billings, & H. Nies (Eds.), *Long-Term Care in Europe. Improving Policy and Practice* (pp. 19–41). Basingstoke: Palgrave Macmillan.

Pfau-Effinger, B. (1998). Gender Cultures and the Gender Arrangement – A Theoretical Framework for Cross-National Comparisons on Gender. *The European Journal of the Social Sciences, 11*, 147–166.

Pfau-Effinger, B. (2005). Culture and Welfare State Policies: Reflections on a Complex Interrelation. *Journal of Social Policy, 34*, 3–20.

Reichert, M. (2012). Vereinbarkeit von Erwerbstätigkeit und Pflege – eine Bestandsaufnahme [Reconciliation of Work and Care – A Review]. In R. Bispinck, G. Bosch, K. Hofemann, & G. Naegele (Eds.), *Sozialpolitik und Sozialstaat* [Social Policy and Welfare State] (pp. 323–333). Wiesbaden: Springer.

Rodrigues, R., Huber, M., & Lamura, G. (2012). *Facts and Figures on Healthy Ageing and Long-Term Care*. Vienna: European Centre.

Rossi Mori, A., & Dandi, R. (2012). *The Influence of Technology on Long-Term Care Systems*. ENEPRI Policy Brief No. 10.

Rostgaard, T., & Pfau-Effinger, B. (2011). *Care Between Work and Welfare in European Societies*. Basingstoke: Palgrave Macmillan.

Ruppe, G. (2011). *Active Ageing and Prevention in the Context of Long-Term Care. Rethinking Concepts and Practices.* European Centre, Policy Brief July 2011. Vienna: European Centre.

Sainsbury, D. (1996). *Gender, Equality and Welfare States*. Cambridge: Cambridge University Press.

Schulmann, K., & Leichsenring, K. (2014). *Social Support and Long-Term Care in EU Care Regimes. Framework Conditions and Initiatives of Social Innovation in an Active Ageing Perspective*. Vienna: European Centre. MoPAct Report, #8.1.

Schulmann, K., & Leichsenring, K. (2015). *A Qualitative Inventory of the Key Drivers of Social Innovation in the Delivery of Social Support and Long-Term Care*. Vienna: European. MoPAct Report, #8.3.

Simonazzi, A. (2009). Care Regimes and National Employment Models. *The Cambridge Journal of Economics, 33*(2), 211–232.

Széman, Z., Ghenta, M., Matei, A., Tróbert, M., Schulze, S., & Paat-Ahi, G. (2016). Mobilising the Social Innovation Potential in LTC: Expanding Services. Policy Paper #4. In K. Leichsenring, K. Schulmann, G. Casanova, G. Ghenţa, G. Lamura, A. Matei, … M. A. Tróbert (Eds.), *Potentials for Social Innovation in Social Support and Long-Term Care* (pp. 4–15). A Series of Policy Papers. Vienna: European Centre. MoPact Report #8.4.

Timonen, V., & Doyle, M. (2007). *Home Care for Ageing Population: A Comparative Analysis of Domiciliary Care in Denmark, The United States and Germany*. Cheltenham: Edward Elgar.

Van der Heide, I., Snoeijs, S., Melchiorre, M. G., Quattrini, S., Boerma, W., Schellevis, F., & Rijken, M. (2015). *Innovating Care for People with Multiple Chronic Condition in Europe.* ICARE4EU Project. Final Report. ICARE4EU. Retrieved from http://www.icare4eu.org/pdf/State-of-the-Art_report_ICARE4EU.pdf

Walker, A., & Foster, L. (2013). Active Ageing: Rhetoric, Theory and Practice. In R. Ervik & T. S. Linden (Eds.), *The Making of Ageing Policy: Theory and Practice in Europe* (pp. 27–52). London: Edward Elgar.

WHO. (2015). *WHO Global Strategy on Integrated People-Centred Health Services.* Interim Report. Retrieved March 20, 2018, from WHO. http://apps.who.int/iris/bitstream/10665/155002/1/WHO_HIS_SDS_2015.6_eng.pdf?ua=1&ua=1

WHO. (2016). *What Is 'Active Ageing'?* Retrieved March 20, 2018, from World Health Organisation (WHO). http://www.who.int/ageing/active_ageing/en/

Williams, F. (2012). Converging Variations in Migrant Care Work in Europe. *Journal of European Social Policy, 22*(4), 363–376.

10

Promoting the Political Inclusion and Participation of Older People: Social, Psychological and Institutional Perspectives

Amílcar Moreira and Mikkel Barslund

This chapter builds primarily on research conducted members of the Mobilising the Potential of Active Ageing in Europe (MOPACT) consortium, namely, *Amílcar Moreira, Rui Costa Lopes, Roberto Falanga, Mariana P. de Miranda, Pedro Alcântara, Luis Manso, Bruno de Carvalho*—University of Lisbon, Institute of Social Sciences (ICS), Lisbon, Portugal; *Mikkel Barslund, Sara V. de Almeida*—Centre for European Policy Studies (CEPS), Brussels, Belgium; *Andrea Principi, Francesco Balducci, Marco Socci, Roberta Papa*—National Institute of Health and Science on Aging (INRCA), Ancona, Italy; *Hannah Swift, Dominc Abrams*—University of Kent (UKENT), Canterbury, UK; *Andreas Cebulla, Dave Wilkinson*—National Institute of Economic and Social Research (NIESR).

A. Moreira (✉)
Institute of Social Sciences, University of Lisbon, Lisbon, Portugal
e-mail: amilcar.moreira@ics.ulisboa.pt; https://www.ics.ulisboa.pt/pessoa/amilcar-moreiraand

M. Barslund
Centre for European Policy Studies, Brussels, Belgium
e-mail: mikkel.barslund@ceps.eu; https://www.ceps.eu/profiles/mikkel-barslund

A. Walker (ed.), *The Future of Ageing in Europe*,
https://doi.org/10.1007/978-981-13-1417-9_10

Introduction

As established in Chap. 1 and revisited throughout this book, Europe is ageing rapidly. The share of the older European population (65+) increased from 15.8% in 2001 to 18.5% in 2014 and is projected to increase further (Eurostat 2015). Moreover, the share of the oldest old, those aged 85 or above, will constitute a larger share of the group of older people (Barslund and von Werder 2016).

The expected changes in the needs and preferences associated with demographic ageing require that we put in place the institutional mechanisms that enable older citizens to express their views during all stages of the policy-making process—from the design of public policies to the delivery of services on the ground (Walker 1999). Besides contributing to a more enhanced participatory democracy, this should facilitate institutional change, reduce the risk of paralysing political conflicts and help to promote social cohesion and quality of life. On the face of this, it is surprising how little we actually know about the civic engagement of older people, particularly compared to what we know about such engagement among young people (Quintelier 2007) or about the participation of older people in other domains of social life (Walker 2018).

This chapter aims to expand our knowledge about the factors that condition the political participation of older people and to identify mechanisms and good practices that might help to improve their civic engagement. It does so by offering an innovative analytical framework that recognises that political behaviours are the product of individual (age, lifecycle and cohort effects), symbolic (social representations of old age) and institutional (blockages and opportunities) factors.

In the first part of the chapter, the fundamental pillars of our analytical framework to study the political participation of older people are laid out together with an assessment of the current state of the art on the key dimensions of this framework. The second part uses this analytical framework to gain a more comprehensive understanding of individual, symbolic and institutional mechanisms that shape older people's involvement in the public sphere. Building on these findings, we conclude by identifying a number of avenues of further research on this topic.

The Political Participation of Older People in Europe: Towards a More Comprehensive Framework of Analysis

Building on insights from demography, sociology, political science and social psychology, in this section we aim to provide a comprehensive framework of analysis that acknowledges that the political participation of older people is influenced by range of mechanisms that act at the individual, symbolic and institutional levels.

Central to this approach, reflecting the arguments for it in Chap. 1, is the idea that the political participation of older people must be analysed from a life-course perspective. This means that older peoples' civic engagement must be understood as (a) being influenced by a series of resources and experiences that individuals accumulate over time and (b) being part of broader life transitions, including the transition to retirement and transitions into an out of care.

Building on this general insight, the analytical framework brings two important contributions to the study of the political participation of older people in Europe. First, it acknowledges that, in addition to individual-level factors such as age, labour market status or income, the decision of older people to participate in the political sphere is influenced by social representations of older people and ageing. The Risks of Ageism Model (Swift et al. 2017) suggests that there are three pathways through which ageism and negative attitudes towards age impact on older people:

- By being a target of age discrimination[1];
- Through stereotype embodiment, which occurs when stereotypes that were once targeted on 'other' older people become relevant to the self. For example, evidence suggests that people who hold more negative stereotypes of older people may also expect worse outcomes from their own experience of ageing (Levy 2009);
- Via stereotype threat, which refers to the threat experienced by an individual when they feel a situation puts them at risk of confirming a negative stereotype about their group.

This implies that, in order to fully understand if, and how individuals participate in the civic sphere, we need to understand how these types of symbolic mechanisms are (re)produced in society, and in what way do they shape individual participation decisions.

Second, our analytical framework acknowledges that the political participation of older people is fundamentally dependent on the institutional opportunities available to them in the political system. As Campbell (2003) shows, even in the context of low participation, given the right context and the set of institutional incentives, older people can be mobilised into participating in the political sphere. This means that in addition to looking at individual and symbolic factors that shape older people participation decisions, it is necessary to take into consideration how these decisions are influenced by the institutional incentives that frame those decisions.

The Political Participation of Older People from a Life-Course Perspective

After an initial period of interest (Crittenden 1963; Nie et al. 1974), the issue of how active citizenship evolves over the lifecycle has remained largely ignored by political scientists and political sociologists. However, with the increasing interest in the impact of ageing in developed economies, a body of evidence has slowly been building in recent years (Goerres 2009; Grasso 2014).

There is a significant level of conceptual and methodological diversity in how this issue is tackled. Some scholars study more intensive forms of institutional participation such as being a member of a political party, a trade union or a social movement organisation (Goerres 2009; Grasso 2014). Others look at less intensive forms of institutional participation, such as voting, giving money to or volunteering for to a political organisation/candidate or volunteering for political organisations/candidates (Burr et al. 2002; Campbell 2002; Smets and Neudorf 2014). In some other cases, scholars compound various forms of political participation into summary measures of active citizenship (Erkulwater 2012).

Notwithstanding the importance of determining how civic involvement is measured, a more crucial issue in studying how political participation evolves over the lifecycle concerns what scholars have come to call the 'age-period-cohort' identification problem. As Goerres (2009) rightly highlights, the relationship between age and political participation is the product of a number of (confounding) effects:

- Age (or lifecycle) effects, which refer to a set of common conditions that impact on the political participation of individuals of a given age or in a given stage of their lifecycle;
- Cohort effects, which refer to a set of common conditions that shaped the political socialisation of individuals of a given demographic cohort, or of a given generation (generational effects), and that will reflect in their political behaviour throughout the rest of their lives;
- Period effects, which refer to the influence of historical context in which measurements of political participation were taken.

Disentangling the effects of age, period and cohort is a well-known empirical problem (Neundorf and Niemi 2014). The common solution to deal with this in the literature on political participation or other forms of active citizenship is to use alternative measurements for age, period and cohort effects (Goerres 2009). More recently, scholars have explored the heuristic potential of multilevel models to deal with the age-period-cohort identification problem. Building on this approach, and on data from the European Election Studies database, Bhatti and Hansen (2012) find that voter turnout at European Parliament elections declines significantly after the mid-1950s. Grasso (2014) finds a similar curvilinear effect of age on party membership and on the probability of signing a petition using the European Values Study. The same picture emerges from both Wass's (2007) study on Finish elections, and Smets and Neudorf's (2014) work on US presidential elections.

Notwithstanding the important contribution that this recent set of studies have made to the study of how political participation evolves over the lifecycle, they are nonetheless marred by the fact that they fail to acknowledge that the influence of age, cohort and period effects on individuals' political participation decisions varies significantly across nations. Moreira et al. (2018)—discussed further below—attempt to move the research forward on this issue.

The Importance of Social Representations

Over recent years, a growing body of evidence has developed regarding the representations of older people in society. Research suggests that older people are not only stereotyped as frail, ill, dependent and incompetent (Marques et al. 2014; Coudin and Alexopoulos 2010), but also wise, experienced and more moral than younger adults (Swift et al. 2012; Abrams et al. 2011). Several studies have explored age-related stereotypes across age and cultural groups, but these tend to focus on old-age stereotypes and show that younger adults have stronger stereotypes and are more likely to endorse negative views of older adults (North and Fiske 2013a).

Many of the negative and positive representations of ageing are captured within the Stereotype Content Model (Fiske et al. 2002). This model proposes that competence and warmth constitute the two fundamental dimensions of social perception, and thus are the basis upon which stereotypes are formed. Thus, social groups can be perceived as high or low on competence or warmth. Depending on how a certain group is perceived, this will elicit a set of corresponding emotions and also behaviours. Thus, a group low on competence and high on warmth will elicit a behaviour of pity-driven protection, whereas a group high on competence, but low on warmth, will elicit a behaviour of competition, driven by feelings of envy. In the particular case of older people, these tend to be perceived as high on warmth, and low on competence.

North and Fiske (2013a, b) have recently made an important contribution to the understanding of ageism. The authors identified a set of prescriptive social norms as to how older people should behave in society, namely, with regard to the succession of desired resources (notably, jobs), the consumption of shared resources (namely, public expenditure) and the adoption of age-adequate lifestyles. Based on this, the authors show that the perceived violation of these prescriptive norms (older people holding on to their jobs not leaving opportunities for youths; older people being a burden to the health care and social security systems; older people entering social spheres that are not suitable for their age) will elicit resentment from younger people—which suggests a shift from the traditional pity-based prejudice towards the old.

With the view to further explore the heuristic potential of North and Fiske's (2013a) framework, below we discuss the results of a set of confirmatory experimental studies as presented by Lopes et al. (2018) and also look at how these prescriptive norms are reproduced through print media (Swift et al. 2018).

From Supply to Demand: Participatory Approaches for the Involvement of Older People in Policy-Making

As mentioned above, in order to get a full grasp of the factors that shape the participation of older people in the civic sphere, we also need to look at the opportunities for participation in the institutional mechanisms of democratic deliberation. Interest in the involvement of citizens in policy-making can be traced back to the experiments of participatory budgets carried out in Latin America at the end of the 1980s (Avritzer 2006). These experiences helped to shift from a vision of citizens as mere 'beneficiaries' of public policies towards new forms of self-determination.

From the 1990s onwards, debates on participatory democracy became more explicitly concerned with effective forms of operationalisation, and emphasis has increasingly been given to the empowerment of citizens (UNDP 1993). Much of this focus built on the seminal work of Arnstein (1969), who argued that the effectiveness of mechanisms for involving citizens in policy-making depends on the level of redistribution of information and resources, and on the possibility to influence decision-making. Building on this, the author put forward a 'ladder of participation', which established a hierarchy of participatory approaches in terms of their ability to enable citizens' influence over decision-making.

Arnstein (1969) argued that despite creating an opportunity for citizens to express their views, lower forms of participation, such as informing or consultation, do not empower citizens with the ability to effectively shape the results of the policy-making process. Effective empowerment only takes place when citizens are allowed to negotiate with powerholders (Partnership) or have the effective power to influence decisions (Delegated Power and Citizen Control).

Arnstein's (1969) seminal study shaped a plethora of typologies of participation, both in academia and in national, and international, policy-making institutions. For instance, Lowndes et al. (2001) have identified five modalities of citizen participation:

- *Consumerist methods*, for example, complaints schemes, satisfaction surveys, opinion polls;
- *Traditional methods*, for example, public meetings, consultation documents, co-option committee, Q&A sessions;
- *Forums*, for example, service user, area/neighbourhood-based, issue-related, shared interest;
- *Consultative methods*, for example, interactive websites, citizens' panels, referendum;
- *Deliberative innovations*, for example, focus group, community plan, visioning exercise, user management, citizens' juries.

Despite the significant attention that the issue of participatory democracy has attracted, there are not many studies that focus specifically on the involvement of older people in policy-making. One exception is Carter and Beresford (2000), who establish a typology of models of involvement of older people in policy-making, covering:

- Advocacy and Information;
- Senior Councils/Forums;
- User Panels;
- Consultation;
- User-groups;
- User-led services
- Direct Payments;
- Networks;
- Campaigning and direct action;
- Initiatives in other countries.

Walker (2006) stresses that the aims and mechanisms for involving older people in policy-making depend on the level at which policies are decided/implemented:

- at the national level, the involvement of older people is designed to leverage local 'person-centred' experiences and to improve the preparation of guidelines and policy-documents (see also Older People's Steering Group 2004). This is traditionally done through 'representative' forms and consensual models of participation;
- at the meso/local level, there is a tendency to use participatory mechanisms to improve the delivery of health and social services.

In a paper discussed in more detail below, Falanga et al. (2018) chart the different approaches adopted in Europe for involving older people in policy-making, and identify best-practices in this field.

Promoting the Political Inclusion and Participation of Older People: New Insights

Building on the analytical framework and related literature laid out in the first part of this chapter, the second half looks at new insights into the political participation of older people generated from the Mobilising the Potential of Active Ageing in Europe (MOPACT) project.

Political Participation Tends to Remain Stable over the Life Course, with Some Crucial Exceptions

As mentioned above, the existing evidence on how political participation evolves over the life-course fails to adequately acknowledge that age, cohort and period effects vary significantly across nations. As we argued, this was in part due to the lack of adequate longitudinal cross-national databases on this topic. In order to overcome this gap in the evidence base, Moreira et al. (2018) used data from the four waves (1981–1984, 1990–1993, 1999–2001) of the European Values Survey[2] to build a pseudo-panel to study how political participation evolves over time (Deaton 1985). The underlying assumption is that individuals in the same cohort share a number of common circumstances in the early stages of their lives, and consequently also share certain attitudes and/or behaviours. This allows us to control for the effect of (time-variant and time-invariant) factors in individual's behaviour over the lifecycle.

Building on Hoskins and Mascherini (2009), Moreira et al. (2018) look at life-course changes in involvement in the following domains of civic life:

- Participation in the party politics, namely, being member of a political party or group;
- Participation in employment-related civil society organisations, namely, trade unions or professional associations;
- Participation in other civil society organisations, namely, environmental or human rights organisations;
- Membership of community organisations, namely, cultural activities, youth or social welfare organisations.

Results suggest that participation in the various fields of the public sphere seems to be relatively stable over the lifecycle. The exception is participation in employment-related civil society organisations, with individuals starting to drop out as they come close to withdrawing from the labour market. This suggests that changes in the overall level of participation in civic life are very much related to changes in individuals' labour market status. The research also shows that changes in individuals' economic status, rather than changes in the family situation, explain variations in the overall level of civic engagement.

With the view to further explore how transitions in individuals' lifecycle impact on their civic engagement, and using data from the SHARE survey,[3] Principi et al. (2018) looked at how both the existence of care responsibilities and transitions in and out of providing care may impact on the participation of older people in the community.[4] The results of a (panel-based) random-effect regression model suggest that providing informal family care outside the household is correlated with higher levels of social engagement. However, this is less true for women who have fewer opportunities than men to be active in other domains of social of civic life.

Cross-National Variations Matter

An issue often ignored when looking at the political participation of older people in Europe is that there are large cross-national differences. Using the 2002 European Social Survey, Barslund et al. (2018) created a set of innovative composite indicators with the view to chart cross-national differences in the civic engagement of older people. Each composite indicator measures the levels of participation in a set of activities in three key areas: representative democracy, community life and protest and social change movements (Hoskins and Mascherini 2009).

In a first instance, countries were ranked by reference to an index that simply measures the levels of participation in the senior community (60+). This reveals that political participation of older people broadly reflects the national patterns of civic engagement. Hence, a simple comparison suggests that older people in Nordic countries are more likely to participate in civic life than those in the Southern (Portugal, Greece) and Eastern (Hungary, Poland) peripheries of the European continent. However, this might obfuscate important cross-national differences in the relative level of civic engagement of older people in Europe. Therefore, Barslund et al. (2018) explored a ranking of countries based on a measure of civic engagement of the senior community *relative* to younger age groups, comparing the difference in the level of political participation of those aged 60+, to that of those aged between 30 and 59.

Once in-country age differences are considered, the level of participation by older people in some of the best-performing cases (Norway, Denmark and Austria) is significantly lower than previously measured. Conversely, the importance of the political participation of older people in Portugal, Hungary, Poland and Italy increases. In other words, the decrease in civic engagement in the latter countries from young to old is much smaller than in the countries of Norway, Denmark and Austria.

These findings are confirmed by further MOPACT research (Moreira et al. 2018). Based on the results of a cross-classified multilevel model where age, cohort and period effects are clustered at the country level, it appears that the importance of cohort membership in explaining variations in civic engagement might be over-estimated and that, more than any other factor, individual differences in political participation are very much shaped by the cross-national variations in social institutions.

Prescriptive Norms About the Role of Older People in Society Impact on Ageing Stereotypes and Behaviours Towards Older People

Building on the seminal study by North and Fiske (2013a, b), Lopes et al. (2018) conducted two comparable experimental studies (in the UK and Portugal[5]) to test if being exposed to the perception of older people as violating the prescriptive norms elicits the emergence of negative stereotypes about older people. In the Portuguese case, the perceived violation of the succession norm is associated with a downgrade of the perception of older peoples' competence. On the other hand, a perceived violation of the identity norm seems to improve the perception of older peoples' competence. No statistically significant effect was found for evaluations of older peoples' warmth, however. In the UK, there is no statistically significant effect of the perceived violation of these norms on how respondents evaluate older persons either in terms of warmth or competence.

This study also suggests that the perceived violation of the aforementioned norms affects individuals' assessment of older peoples' contribution to the economy. In the Portuguese experimental study, the results of a MANOVA analysis suggest that—compared to those in the control group—individuals that are exposed to the image of older people violating these prescriptive norms are more likely to perceive them as making a negative contribution to the economy. However, the UK experimental study fails to confirm this finding. Curiously, both the Portuguese and the UK studies fail to find any statistically significant evidence that the perception of norm violations is associated with perceiving older people as posing a symbolic threat to younger generations.

In addition to the potential impact on individual attitudes, this research suggests that the perceived violation of these prescriptive norms may elicit the emergence of discriminatory behaviour, namely, in what concerns the use of public spending in the field of education. Thus, both in Portugal and the UK, experiment participants faced with a perceived violation of the succession norm are more likely to support spending on the education of younger people. The particular sensitivity of policy preferences for public spending in education is further reinforced by the fact that, in the Portuguese study the perceived violation of the succession, consumption or identity norms does not have a statistically significant impact on other individuals preference concerning spending in pensions, unemployment protection or health care.

Print Media Plays an Important Role in the (Re)production of Prescriptive Norms About the Role of Older People in Society

With the view to further explore of how ageing-related prescriptive norms have an impact on attitudes towards older people, Swift et al. (2018) have carried out an analysis of a representative sample of newspaper articles published in random weekdays in the UK and Portugal in the period between January 1st of 2013 and 30th of June in that year,[6] which shows that print media help to (re)produce prescriptive norms about how older people should behave in society—even if depictions of succession, consumption or identify norms are not as salient as are more general representations of older people in terms of warmth and competence (North and Fiske 2013a, b). Moreover, it was found that prescriptive norms about how older people should behave in society are more salient in UK print media than in Portugal. Third, and also importantly, this study shows that, both in the UK and in Portuguese print media, the consumption norm is much more salient than the other prescriptive norms.

In addition to charting the salience of prescriptive norms about older people in print media, the research also found that the salience of these norms varies by reference to the newspapers' genre and/or political orientation. Thus, the consumption norm is more salient in broadsheet newspapers than in tabloids. On the other hand, the succession and identity norms are more salient in left/centre-left leaning newspapers. A possible explanation for this is that as tabloids and right/centre-right leaning newspapers tend to target senior readers as a key audience, and therefore avoid bringing about normative challenges to their role in society. This would suggest that, rather than genre or political orientation, the target audience of the newspaper might be a more useful variable to explain the salience of norms about the role of older people in society.

Policy Makers Tend to Favour a More Consultative Approach When Involving Older People in Policy-Making

Falanga et al. (2018) conducted an extensive search of current (or recent) programmes/initiatives aimed at involving the participation of older people in aspects of policy-making in Europe. Of a total 80 programmes/initiatives found, a sample of 37 was selected for which there was sufficient information that allowed for a systematic comparison. Of the 37 initiatives, 28 were consultative decision-making practices, 6 consultative policy implementation practices, 2 co-decisional decision-making practices and 1 co-decisional policy implementation practice. These programmes/initiatives were then grouped according to where they are positioned in the policy-making cycle (decision making or policy implementation) and the type of participatory approach adopted (consultative or co-decisional). As can be seen in Fig. 10.1, mechanisms of co-decision involving older people are rare. Most of the cases identified consist of consultative bodies, such as Older Peoples' Councils or Older Peoples' Forums, mostly at the decision-making level. The use of consultative bodies for involving older people in the actual delivery of services is also rare. Finally, these consultative bodies tend to involve individual (elected or nominated) delegates or representative organisations. The direct involvement of individuals from the local community is unusual.

Guidelines for Enhancing the Participation of Older People in Policy-Making

With a view to identifying the factors that can enhance (or hinder) the participation of older people in policy-making, Falanga et al. (2018) identified a set of best practices,[7] which were then the object of a set of 'peer review' workshops. Building on the discussions during the 'peer

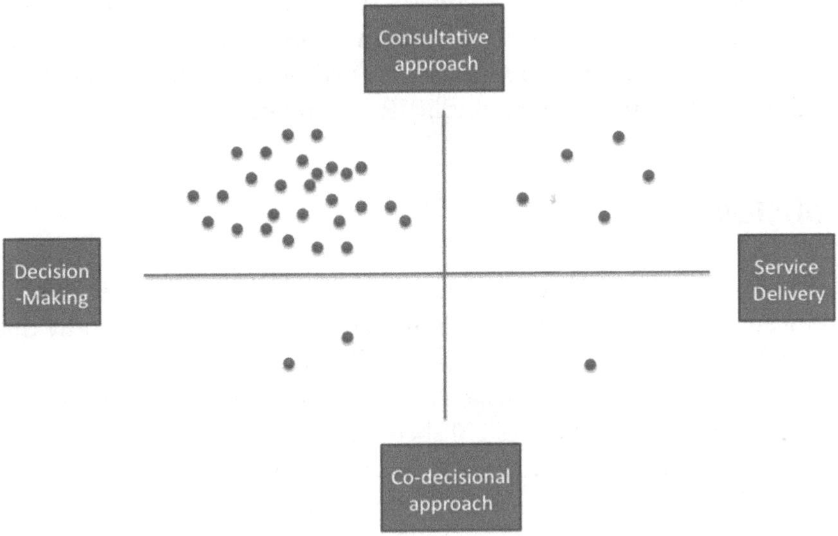

Fig. 10.1 A typology of approaches of the involvement older people in policy-making in Europe. (Source: Falanga et al. 2018)

review' workshops, the authors identified a set of guidelines that policy-makers should consider when setting-up initiatives to promote the participation of older people in policy-making:

- *Commitment is Paramount.* Participation needs public statements of commitment by political authorities and civil society.
- *Beware of Diversity.* Political authorities should take into consideration the ways older people from different social status—namely, those with lower qualifications—and different ages—namely, those aged 80+—can have access and participate.
- *Invest in Training.* The agencies in charge for the organisation of participatory initiatives should invest on training activities for the actors involved in the process.
- *Acknowledge the Benefit of Good Communication.* Civil society should have the chance to be informed about the state of the art of the policies to be debated, and should also be regularly updated about the advancement of the participatory process.

- *Be Comprehensive in Your Approach.* Political authorities should make sure that participatory initiatives are sustained by broader institutional changes aimed at overcoming entrenched bureaucratic processes and addressed to improve open public data systems.

Conclusion

Building on an innovative analytical framework that recognises that the political participation decisions are influenced by a number of individual (age, lifecycle and cohort effects), symbolic (social representations of old age) and institutional (blockages and opportunities) mechanisms, this chapter aims to expand the established knowledge about the factors that condition the political participation of seniors in Europe, and to identify mechanisms and good practices that might help to improve their civic engagement. This final section summarises the key findings produced during the MOPACT project, and reflects on how they should shape future research on this topic.

The first key finding is that the transition to retirement is associated with a decrease in the level of civic engagement. However, rather than a withdrawal from public life, this merely reflects the disengagement of employment-related civil society organisations as individuals drop out of the labour market. Participation in other domains of civic life seems to remain stable over the life course, and that transitions into care do not seem to have an impact on participation decisions. Furthermore, MOPACT researchers found that cohort membership in shaping political participation decisions might be over-estimated in the literature. Once we model cohort effects as depending on national institutions/contexts, their importance in explaining individual variations in civic engagement drops significantly. These findings would suggest that more research is needed the role of national social and economic institutions in moderating the influence of age, period and cohort membership in individuals' political participation decisions.

Second, MOPACT researchers uncovered evidence that perceived violations of the consumption, succession and identity norms seem to strengthen negative evaluations concerning older peoples' competence and their contribution to the economy. Moreover, perceived violations of the succession norm increase the likelihood of being supportive of spending on education, which favours the younger generations. Print media do play a role in the reproduction of prescriptive norms about older people and ageing—albeit less than we would have anticipated. The consumption norm is much more salient than the other prescriptive norms. These findings strengthen our argument that further research into this topic should take into consideration symbolic mechanisms in individuals' civic engagement. In particular, further research should go into how these norms actually impact on political participation. Moreover, more evidence should go into the role of the media (in its different formats) in (re) producing prescriptive norms about the role of older people in society. Finally, future research should explore the potential role of the 'active ageing' policy paradigm as a source for an alternative normative standpoint by which society evaluates the role of older people in society (see Chaps. 1 and 11).

Third, MOPACT research has shown that the involvement of older people in policy-making in Europe takes place fundamentally at the decision-making level, and mostly on a consultative basis. The involvement of older people in decisions concerning the implementation of services is, on the other hand, rare. In light of these findings, we believe further research should go into the mechanisms/initiatives designed to promote the participation of older people in policy-making. Thus, special attention should be given to the mechanisms that facilitate/hinder the participation of from certain social groups, namely, institutionalised persons, those with severe mobility limitations or those with very low levels of education. Moreover, more research is needed into the effectiveness of the participatory mechanism/initiatives in shaping policy-making, namely, in to what shapes the delivery of services on the ground.

Notes

1. Age discrimination is a social process by which older people can be excluded or denied access to a product, service or treatment (Abrams 2010), such as employment or in health and social care, and can be direct

or indirect in nature. Direct age discrimination occurs when someone treats another less favourably because of their age, whereas indirect age discrimination happens when a social custom, policy or established practice puts someone at a disadvantage because of their age.

2. The European Values Survey (EVS) is a large-scale, cross-national, survey that since 1981 has been tracking changes in attitudes, social values and behaviours of respondents in Europe, US, Canada and, more recently, Russia and Turkey.
3. Wave 1 (2004) and Wave 2 (2006–2007) from the Survey of Health, Ageing and Retirement in Europe (SHARE).
4. This is captured by a composite index that measures the (cumulative) engagement in three key dimensions of public life: social participation (taken part in religious organisations; attend educational or training courses; go to sport, social or other kind of clubs); done voluntary or charity work; or taken part in a political or community-related organisations.
5. The Portuguese experiment involved 115 undergraduate Psychology students at a Portuguese university. The UK experiment involved 207 respondents from Prolific Academic, an online crowdsourcing platform for conducting research. Participants were randomly assigned to one of the three experimental conditions—consumption threat, succession threat, identity threat or to a control condition.
6. In order to secure a representative sample of articles in each of these countries, newspapers that reflect the structure of the print media market were selected, by reference to genre (tabloid vs. broadsheet), political orientation (left/centre left; centre/no clear political orientation; centre-right/right) and influence, as measured by the level of sales. In the UK, the selection of newspaper also took into consideration the geographical coverage of the newspapers. In the end, the UK sample was built using articles from five newspapers (The Times; The Guardian; The Mirror; Daily Mail; and the Financial Times), while the Portugal sample was extracted from three newspapers (Correio da Manhã; Jornal de Negócios and Público). This has produced a total sample of 1234 articles—738 from the Portuguese subsample, and 469 from the UK subsample.
7. The following best-practices in the involvement of older people in policy-making were identified: Senior Citizens' Councils (SSC), in Denmark; the Partnership for Older People Programme (POPP), in Dorset (UK); and the Senior Citizens' Participatory Budget, in Alfândega da Fé (Portugal).

References

Abrams, D. (2010). *Processes of Prejudices: Theory, Evidence and Intervention, Equalities and Human Rights Commission.* Retrieved from http://www.equalityhumanrights.com/uploaded_files/research/56_processes_of_prejudice.pdf

Abrams, D., Russell, P., Vauclair, M., & Swift, H. (2011). *Ageism in Europe: Findings from the European Social Survey. Age UK.* Retrieved from http://www.ageuk.org.uk/documents/en-gb/id10704%20ageism%20across%20europe%20chapter%20interactive.pdf

Arnstein, S. (1969). A Ladder of Citizen Participation. *Journal of the Royal Town Planning Institute, 1*(6), 216–224.

Avritzer, L. (2006). New Public Spheres in Brazil: Local Democracy and Deliberative Politics. *International Journal of Urban and Regional Research, 30*(6), 623–637.

Barslund, M., & von Werder, M. (2016). Measuring Dependency Ratios Using National Transfer Accounts. *Vienna Yearbook of Population Research 2016, 14*, 155–185.

Barslund, M., Almeida, S. V., & Principi, A. (2018). Measuring and Cross-Country Comparisons of Active Citizenship Among Different Generations in the EU. In *Centre for European Policy Studies.* Belgium: CEPS.

Bhatti, Y., & Hansen, K. M. (2012). The Effect of Generation and Age on Turnout to the European Parliament: How Turnout Will Continue to Decline in the Future. *Electoral Studies, 31*(2), 262–272.

Burr, J. A., Caro, F. G., & Moorhead, J. (2002). Productive Aging and Civic Participation. *Journal of Aging Studies, 16*, 87–105.

Campbell, A. L. (2002). Self-Interest, Social Security, and the Distinctive Participation Patterns of Senior Citizens. *The American Political Science Review, 96*(3), 565–574.

Campbell, A. L. (2003). Participatory Reactions to Policy Threats: Senior Citizens and the Defense of Social Security and Medicare. *Political Behavior, 25*(1), 29–49.

Carter, T., & Beresford, P. (2000). *Age and Change: Models of Involvement for Older People.* York: Joseph Rowntree Foundation.

Coudin, G., & Alexopoulos, T. (2010). "Help me! I'm old!" How Negative Aging Stereotypes Create Dependency Among Older Adults. *Aging and Mental Health, 14*(5), 516–523. https://doi.org/10.1080/13607861003713182.

Crittenden, J. (1963). Aging and Political Participation. *The Western Political Quarterly, 16*(2), 323–331.

Deaton, A. (1985). Panel Data from Time Series of Cross-Sections. *Journal of Econometrics, 30*(1), 109–126.

Erkulwater, J. L. (2012). Political Participation Over the Life Cycle. In K. L. Schlozman, S. Verba, & H. E. Brady (Eds.), *The Unheavenly Chorus: Unequal Political Voice and the Broken Promise of American Democracy* (pp. 199–231). Princeton: Princeton University Press.

Eurostat. (2015). Population*: Structure Indicators. Proportion of the Population Aged 65 Years or More*. Retrieved March 20, 2018, from Eurostat: http://appsso.eurostat.ec.europa.eu/nui/submitViewTableAction.do

Falanga, R., Cebulla, A., Socci, M., & Moreira, A. (2018). *The Participation of Senior Citizens in Policy Making: Developing a Taxonomy.* University of Lisbon: Institute of Social Sciences.

Fiske, S., Cuddy, A. J., Glick, P., & Xu, J. (2002). A Model of (Often Mixed) Stereotype Content: Competence and Warmth Respectively Follow from Perceived Status and Competition. *Journal of Personality and Social Psychology, 82*(6), 878.

Goerres, A. (2009). *The Political Participation of Older People in Europe: The Greying of Our Democracies.* Basingstoke: Palgrave Macmillan.

Grasso, M. T. (2014). Age, Period and Cohort Analysis in a Comparative Context: Political Generations and Political Participation Repertoires in Western Europe. *Electoral Studies, 33*, 63–76.

Hoskins, B., & Mascherini, M. (2009). Measuring Active Citizenship Through the Development of a Composite Indicator. *Social Indicators Research, 90*(3), 459–488.

Levy, B. R. (2009). Stereotype Embodiment: A Psychosocial Approach to Aging. *Current Directions in Psychological Science, 18*, 332–336.

Lopes, R. C., Miranda, M. P., Swift, H. J., Abrams, D., & Moreira, A. (2018). *The Impact of Perceived Threats on Attitudes Towards Senior Citizens: An Experimental Approach.* University of Lisbon: Institute of Social Sciences.

Lowndes, V., Pratchett, L., & Stoker, G. (2001). Trends in Public Participation: Part 1 – Local Government Perspectives. *Public Administration, 79*(1), 205–222.

Marques, S., Lima, M., Abrams, D., & Swift, H. (2014). Will to Live in Older People's Medical Decisions: Immediate and Delayed Effects of Aging Stereotypes. *Journal of Applied Psychology, 44*, 399–408. https://doi.org/10.1111/jasp.12231.

Moreira, A., Manso, L., & Barslund, M. (2018). *Active Citizenship over the Life-Cycle. A Pseudo-Panel Approach.* University of Lisbon: Institute of Social Sciences.

Neundorf, A., & Niemi, R. G. (2014). Beyond Political Socialization: New Approaches to Age, Period, Cohort Analysis. *Electoral Studies, 33*(1), 1–6.

Nie, N. H., Verba, S., & Kim, J. (1974). Political Participation and the Life Cycle. *Comparative Politics, 6*(3), 319–340.

North, M., & Fiske, S. (2013a). Act Your (Old) Age: Prescriptive, Ageist Biases Over Succession. *Consumption, and Identity, Personality and Social Psychology Bulletin, 39*(6), 720–734.

North, M., & Fiske, S. T. (2013b). A Prescriptive, Intergenerational-Tension Ageism Scale: Succession, Identity, and Consumption (SIC). *Psychological Assessment, 25*(3), 706–713.

Older People's Steering Group. (2004). *Older People Shaping Policy and Practice.* York: Joseph Rowntree Foundation.

Principi, A., Balducc, F., Papa, R. M., & Barslund, M. (2018). *Understanding the Relationship Between Informal Family Elder Caregiving and Active Citizenship of Older European People.* The National Institute on Health and Ageing (INRCA – Istituto Nazionale di Riposo e Cura per Anziani).

Quintelier, E. (2007). Differences in Political Participation Between Young and Old People. *Contemporary Politics, 13*(2), 165–180.

Smets, K., & Neundorf, A. (2014). The Hierarchies of Age-Period- Cohort Research: Political Context and the Development of Generational Turnout Patterns. *Electoral Studies, 33*, 41–51.

Swift, H., Abrams, D., & Marques, S. (2012). Threat or Boost: Social Comparison Affects Older People's Performance Differently Depending on Task Domain. *Journals of Gerontology, Series B, 68*, 23–30. https://doi.org/10.1093/geronb/gbs044.

Swift, H., Abrams, D., Lamont, R., & Drury, L. (2017). The Risks of Ageism Model: How Ageism and Negative Attitudes Toward Age Can Be a Barrier to Active Aging. *Social Issues And Policy Review, 11*, 195–231. https://doi.org/10.1111/sipr.1203.

Swift, H. J., Lopes, R. C., Carvalho, B., & Moreira, A. (2018). *Putting Seniors in Their Place: Prescriptive Norms About Seniors in the Portuguese and UK Print Media.* Kent: University of Kent.

UNDP. (1993). *Human Development.* Oxford: Oxford University Press.

Walker, A. (1999). Political Participation and Representation of Older People in Europe. In A. Walker & G. Naegele (Eds.), *The Politics of Old Age in Europe* (pp. 7–24). Buckingham: Oxford University Press.

Walker, A. (2006). Aging and Politics: An International Perspective. In R. Binstock & L. George (Eds.), *Handbook of Aging and the Social Sciences* (6th ed., pp. 339–359). Cambridge, MA: Academic Press.

Walker, A. (2018). *The Future of Ageing in Europe. Making an Asset of Longevity.* Basingstoke: Palgrave Macmillan.

Wass, H. (2007). The Effects of Age, Generation and Period on Turnout in Finland – 1975–2003. *Electoral Studies, 26*, 648–659.

11

Conclusion: Realising Active Ageing

Alan Walker

Introduction

Our journey through the major policy implications associated with ageing Europe now draws to a conclusion. The remaining tasks for this chapter are, first, to summarise the key findings of the research and analyses presented in detail in the previous ones. Second, there is a summary of the major policy reforms necessary to ensure that demographic ageing is a positive experience for individuals and society alike. Then the chapter revisits the central concept of active ageing to explain why it is not at present the dynamic engine it should be in driving a new approach to later life. Finally, in rounding-off the book, there is a very brief resume of the prerequisites for realising active ageing in Europe.

A. Walker (✉)
Department of Sociological Studies, University of Sheffield, Sheffield, UK
e-mail: a.c.walker@sheffield.ac.uk; https://www.sheffield.ac.uk/socstudies/staff/staff-profiles/walker

A. Walker (ed.), *The Future of Ageing in Europe*,
https://doi.org/10.1007/978-981-13-1417-9_11

Key Findings

Strategies of Active Ageing in Europe

Chapter 2 distilled some of the main findings from the whole Mobilising the Potential of Active Ageing in Europe (MOPACT) project:

- Active ageing has been rising across the European Union (EU) Member States in the recent past. This has brought improvements in quality of life among older people as well as tangible economic benefits. Although for most countries there have been improvements for both men and women, there are still significant gender gaps that disadvantage women.
- The greatest challenges in mobilising the strategy of active ageing to benefit individuals and society are faced by Central and Eastern European countries, where actions in this area are typically low. This reflects their weaker institutional capacity and enabling environments for ageing, as well as their developmental stage in the EU. In particular, social participation among older people is very low. These countries with relatively low active ageing outcomes also display higher levels of inequality, especially gender inequality.
- The increasing proportion of the very elderly (80 plus) living alone and the growing gap between life expectancy (LE) and healthy life expectancy (HLE) imply that the future risk of social exclusion will increase unless preventative measures are taken (Chaps. 3 and 6).
- Contrary to popular expectation and widespread policymaker assumptions, ageing is not inevitably linked to the development of chronic diseases. Only 20–25 per cent of HLE is predetermined by genes. In fact, age-associated molecular and cellular damage can be prevented or greatly delayed by lifestyle interventions and enabling environments (Chap. 7).
- The EU has played an important role in supporting research and social innovation but more is required. The funding for the MOPACT project has generated powerful scientific evidence to inform policy priorities for active ageing in Europe. Strong political commitments from national and EU-level policymakers are now required for social policy reforms aimed at realising active ageing. Further funding is required

for similar multidisciplinary projects and for translational work which turns research into policy and practice guidance.

Chapter 2 also undertook the first cluster analysis of EU Member States using the Active Ageing Index (AAI). The key lesson from this research is the need for tailor-made approaches to active ageing, which fit the current stage of development of ageing policies in each country and the varying national institutional structures. For example, as pointed out in Chap. 2, the low scoring countries of Central and Eastern Europe need to take action across a broad front—improving pension adequacy, raising employment rates, and improving opportunities for social and political participation—to enhance the prospects for active ageing. In contrast, the high scoring Nordic and Anglo-Saxon countries require more pin-pointed actions, such as improved employment opportunities in Ireland and higher incomes in old age in the United Kingdom. Although it is difficult to select a single *policy priority* from an overview chapter, this need for sensitivity in the application of active ageing strategies is crucial, coupled with the urgency of the action required in the Central and Eastern European states, if the goal of the EIPAHA is to be achieved.

Economic Sustainability

Chapter 3 conducted a thorough analysis of the economic implications of population ageing and reversed the usually gloomy prognoses from such discussions. Selected key findings are:

- The need to inform citizens about the increased economic risks of living alone and longer than expected.
- Member States need to plan for the rise in the numbers of older people who will live in one-person households.
- The economic challenge of population ageing has been overstated, but substantial increases in the length of working lives are necessary to address the consequences for public finances.
- Linking retirement/pension ages to LE leads to a significant extension of working lives. Contrary to some projections such a policy may be

beneficial in reducing lifetime welfare inequalities, at least in countries with relatively comprehensive social security systems.

The major *policy priority* arising from this analysis is to adjust public pensions to changes in longevity, so that the ratio of working years is maintained on countries with high employment rates for older workers and increased in other countries.

Improving Private Pensions and Retirement Planning

While Chap. 3 considered public pensions, Chap. 4 focused on how to make the private pension sector more effective. Key findings include:

- Private pensions are needed to supplement public pay-as-you go systems.
- Funded private retirement savings can help to create a deeper, more integrated European capital market, thereby enhancing macroeconomic stability and economic growth.
- The design of such pensions can be improved by their customisation to individual needs, by optimising risk sharing over the life course, and encouraging EU-wide portability.
- Information campaigns or financial education programmes are not sufficient to create efficient private retirement planning. The development of choice architecture is essential; so that products are offered in ways that assist consumers to make the optimum choices.

The key *policy priority* from Chap. 4 is the need for governments to regulate the private pension market to ensure that well-designed products, which customise retirement income to individual need, are provided as an industry standard.

Extending Working Lives

Chapter 5 focused on the major issue of the ageing workforce, which follows general population ageing, and specifically the policy goal of all EU governments to extend working lives. The key findings of this investigation include:

- Policies which are primarily aimed at encouraging the extension of working lives by financial incentives are of very limited use unless they are strategically linked to workplace-related human resources policies.
- As with the general case of active ageing policies (Chap. 2), EU countries need to be sub-divided into different groups in determining the most appropriate policy measures to apply. For example, in the Central and Eastern European countries, which lag behind Western Europe ones in the extension of working life, it is external incentives (such as legal regulations and professional advice) that should be the main priority. In the frontrunner countries, such as Germany, it is health, skills training, and working conditions that should be the main focus of attention.
- Within organisations, public and private, the concepts of workability and employability should be the evidence-based reference points for designing policies and practical measures. The pursuit of age management and life course-oriented human resources policies should be adapted to these concepts.
- National older worker programmes (e.g. in Finland or Poland) have proved to be important in raising awareness of the need for policy action in this area. The same is true for information campaigns, particularly in the 'laggard' countries. Active labour market approaches and wage subsidies have been successful in integrating older workers in employment.
- Of rising importance, in terms of extending working lives, are employment risks for older employees in the service sectors. For example, mental health problems are increasing in some sectors in the 'front-runner' countries and with reference to 'work 4.0' (digitalised work).
- The issue of lifelong learning is of increasing importance across Europe but, among the laggard countries, there appears to be huge unmet need and demand.

The key *policy priority* from Chap. 5 is the urgent need to join-up national policies on extending working lives with policies and actions at the workplace.

Participation, Health, and Well-Being

The key findings from Chap. 6, on social engagement and health and well-being, include the following:

- While LE at ages 50, 65, and 85 increased significantly between 2007 and 2015, this was not accompanied by a rise in healthy life years (HLY): LE without activity limitations for both men and women remained stable.
- On average the difference between LE and HLY at age 50 between 2007 and 2015 was 12.2 years for men and 17.2 years for women, which shows that there are still a lot of years of good health to be gained. The four countries in the EU with the largest difference between LE and HLY for both men and women were Slovakia, Estonia, Germany, and Italy.
- Social participation is lower among older people in poor health than those in good health. Yet the determinants of social participation are generally the same in both groups. Therefore, similar measures to promote participation may be recommended for both of them. However, those in poor health face additional barriers to participation, such as depressed mood and lack of transportation.
- In focussing heavily on the labour market in their drive to extend working lives, (Chap. 5) policymakers neglect the fact that different forms of activity compete for time. It is important to facilitate the combination of paid work with other activities, such as volunteering and informal caregiving.
- An increase in social participation levels makes a very limited contribution to HLE (0.0–0.1 years) and therefore will not contribute significantly to the EIPAHA goal of raising HLE by two years by 2020. Despite this, social participation is valuable for individual quality of life and for society, so it is important that governments facilitate such participation, for older people in both good and poor health.

The key *policy priority* arising from Chap. 6 is the need to prevent disability: while chronic diseases will continue this does not mean the end of a satisfying life. However, this becomes more difficult to achieve in the face of increasing disability.

Staying Healthy in Old Age

Chapter 7 represented the concerted inputs to the MOPACT project by biogerontologists. Their scientific research provides much of the evidence base upon which policies and practical interventions to ensure active ageing can be founded:

- Accumulating evidence from experimental studies shows that ageing is not inevitably linked with the development of chronic diseases.
- Only 20–25 per cent of HLE is predetermined by genes: lifestyle and environment play the major role.
- Age-associated accumulation of molecular and cellular damage can be prevented or greatly delayed by lifestyle interventions, such as dietary manipulations.
- Prevention prior to the development of age-associated morbidities is the key to staying healthy in old age. Classic preventative strategies (nutrition, exercise, vaccination) need consistent application and broad public communication. For example, vaccination is a well-established strategy to prevent infectious diseases which account for one-third of mortality in people aged 65 and over, and are a major factor in the development of morbidity and disability.
- Novel preventative strategies, such as dietary interventions, new drugs and stem cells, need successful translation from the understanding of molecular mechanisms, to animal models, to clinic.
- To be successful interventions need to be started early in life and within a life course perspective.

The key *policy priority* from Chap. 7 is the need to implement healthy lifestyle interventions, in particular regarding nutrition and vaccination, across the life course.

Built and Technological Environments

Chapter 8 reported on the broad body of MOPACT research concerned with innovative approaches to raising the health and quality of life of (older) people, which support the goal of active ageing and have the potential to foster economic growth. Key findings include:

- There are huge variations both within and between EU countries in the housing field: the state of dwellings and their accessibility; the local infrastructure; ownership status; and the financial resources of the owner/tenant.
- The key dimensions of actions to enable housing to foster active ageing are active/ambient assisted living, shared housing, accessibility, and ageing friendly neighbourhoods.
- Information and communication technology (ICT)-based assistive technologies within the fields of mobility, housing, and health have a huge potential to support active ageing and offer many productive opportunities in the emerging 'silver economy'.
- With regard to assistive technologies for older persons, European countries face large deployment gaps—the mismatch between R&D efforts on the one hand and both policymakers' expectations and take up on the other.
- To unlock the potential of innovative ICT-based supports, several critical barriers need to be addressed: especially lack of awareness and acceptance among potential users who lack financial support and evidence of efficacy.
- There are many good practices and social innovations in Europe that can help to overcome these barriers, in particular with reference to the issue of user acceptance.
- Differences among the European countries in terms of ICT-take up and literacy/skills provide the opportunity for policy learning and knowledge transfer, but also point to the need for more country-specific approaches. The MOPACT website lists examples of good practice.

The key *policy priority* proposed by Chap. 8 is to urgently plug the deployment gap by creating a web-platform to encourage business-user collaboration. It is also important to focus on the needs of disadvantaged older people.

Long-Term Care

Chapter 9 focused on long-term care (LTC) and considered the potential for active ageing by means of social innovation—as noted in Chap. 1, a unique combination. Key findings include:

- There is an urgent need and a strong potential for innovation in four specific areas within the LTC field: expansion of services; coordination and integration of services; implementation of community-based care; and in employment and professional training.
- There is a clear divide in terms of the presence of institutional structures or framework conditions for promoting innovation within these four areas of LTC. As in several other sectors of ageing policy (Chaps. 2, 5, and 7), there were front-runner countries (such as Denmark, the Netherlands, and Austria) and laggard ones (such as Hungary, Romania, and Estonia) in the application and scaling-up of social innovations. In the first group, innovations tended to be more complex and partly integrated whereas, in the second one, they were less complex, smaller scale, often without public support and single site, isolated interventions.
- As with other analyses in this book, Chap. 9 argued that EU institutions have an important role to play in supporting Member States by fostering the transfer of knowledge between countries, and in promoting national frameworks for improving LTC by embedding incentives for innovation that take a long-term perspective.
- Chapter 9 also makes extensive detailed policy recommendations for each of the four areas of LTC considered. These include:

  - A better definition of target groups in need of LTC.
  - Making informal carers a specific target for support.
  - Responding to the rising need to identify the specific needs of foreign-born older people.
  - Systematic evaluation and scaling-up strategies for innovation programmes that receive public funding.
  - Service planning that is based on needs assessments involving users from within the community.

- A clear focus on 'the local', for example, by activating and building on local resources and existing networks to promote the development of dedicated LTC structures.
- Extending the scope of LTC and integration by addressing new societal challenges in the community towards general issues related to ageing and intergenerational exchange, including ICT applications, built environment, transport, and employment.
- In the area of employment and professional training, the extensive recommendations for policymakers include shifting their focus from acute care to prevention, improving working conditions in LTC, promoting large-scale programmes to train and re-train staff, and revamping the public image of LTC.

The key *policy priority* in LTC is to recognise that innovations and investment in LTC constitute an asset for ageing societies by contributing to employment, reducing poverty and inequality, strengthening intergenerational balance, and eventually leading to efficiency gains in public expenditure on health and social care.

Political Inclusion and Participation

Chapter 10 covered the political inclusion and participation of older people and argued that the changes in needs and preferences brought about by demographic ageing require new institutional mechanisms that enable older people to express their views during all stages of the policymaking process. Key findings include:

- The transition to retirement is associated with a decrease in the level of civic engagement. But, rather than a withdrawal from public life, this is a disengagement from employment-related civil society organisations. Participation in other domains appears to remain stable.
- Perceived violations of the norms of consumption, succession, and identity appear to strengthen negative evaluations among the public concerning older people's competence and economic contribution.

Print media do play a role in the reproduction of prescriptive norms about ageing and older people, although less than anticipated.
- The involvement of older people in policymaking in European countries currently takes place only at the decision-making level and mainly on a consultative basis. Their involvement in decisions concerning the development and implementation of services is rare.

The key *policy priority* arising from Chap. 10 is the need to establish local advisory bodies to represent the views of older people on key issues such as primary health care, LTC, transport, and urban environments and mobility.

Major Policy Priorities

The above catalogue of key findings and recommendations underlines the wealth of knowledge in the MOPACT project ready to be put to good use by policymakers, practitioners, and product producers. Chapter 2 has already pinpointed the top five policy goals arising from this huge body of work and the painstaking efforts of the MOPACT researchers to ensure their research is highly relevant to the challenges associated with ageing in Europe, so only a brief reminder is required here.

Goal 1: Ensuring Economic and Social Sustainability

Key policy actions:

- Linkage of retirement/pension ages to LE.
- Coordination of public policy with workplace human resources policies in the extension of working lives.
- Greatly increased availability of lifelong learning and training.
- Provision of specific guidance with regard to private savings.
- All policy reforms towards financial sustainability must ensure the priority of pension income adequacy.

Goal 2: Promoting Health and Well-Being in Later Life

Key policy actions:

- Urgent actions are required in countries where the gap between LE and HLE is largest: Estonia, France, Germany, Portugal, Slovakia, and Spain.
- Classic public health interventions are required in all EU Member States to promote healthier later lives, in particular promoting nutritional diets, exercise, and vaccination.
- Early life interventions are required to increase HLE.

Goal 3: Provision of Adequate Person-Centred Care

Key policy actions:

- Expansion of existing LTC services (in most countries).
- Greatly increase coordination and integration between health and social care services.
- Innovative design and expansion of community-based care.
- Much greater attention than at present to the training and re-training of LTC staff at all levels.

Goal 4: Adoption of Modern Technologies

Key policy actions:

- Urgent need to close the deployment gap between technological innovation and take up.
- This necessarily entails paying attention to the major barriers to effective deployment, especially lack of awareness among potential users.
- Public subsidies are required to encourage the take-up of ICT-based assistive technologies.

Goal 5: Increased Engagement of Older People in Social, Cultural and Civic Activities

Key policy actions:

- Ensuring that active ageing policies pay close attention to the promotion of social engagement and cultural and civic activities in later life.
- Commitments between political authorities and civil society to encourage greater participation among older people, which should include the creation of local advisory boards to represent the views of older people.
- The heterogeneity of older people (age, education, gender, class, culture, race) has to be taken into account if participation initiatives are to be successful.

Re-focusing Active Ageing Policy

Active ageing is firmly established as the leading global policy strategy response to population ageing, including within Europe. In practice, however, the term 'active ageing' provides a convenient slogan for a wide range of policy discourses and initiatives concerning the impact of demographic change only some of which are worthy of the label. In particular, a productivist ideology which reduces active ageing to working longer has come to dominate these discourses. It is important to understand the reasons for this reduction and demonstrate the much broader potential of active ageing, a potential which, if realised, would more successfully achieve policymakers' limited goal of extending working lives.

Active Ageing in Europe

Following the launch of the idea of active ageing by the World Health Organisation (WHO) (2002), the concept spread rapidly, a process which was assisted greatly by its adoption by the United Nations (UN) in the second global strategy on ageing, the Madrid International Plan of

Action on Ageing (UN 2002). Both the UN's Economic Council of Europe and the EU were quick to seize upon the concept and translate it into policy proposals. Indeed, the EU may be said to have played an important formative role in the development of the concept, for example, in proclaiming the role of older people as active participants in society as a core theme of the 1993 European Year of Older People, and through the activities it sponsored during the 1999 UN Year of Older People (European Commission 1999).

Despite these contributions to the development of active ageing as a policy instrument and its continued application in EU discourses and major events, such as the 2012 European Year of Active Ageing and Solidarity Between Generations, in practice the issue that tends to dominate such initiatives is not active ageing but employment. This productivist orientation became apparent almost from the start even though it contradicted the more comprehensive approach formulated in 1999. When the 2000 Lisbon Council established its ten-year strategy to make Europe 'the most competitive and dynamic knowledge-based economy in the world', it made full employment an overarching goal. Since this goal could not be achieved without raising the employment rate of older (55+) workers in most Member States, it was not surprising that the 2001 Stockholm Summit introduced a specific target employment rate for older workers of 50 per cent. This target was backed-up by a guideline on active ageing which placed the employment of older workers at centre stage in EU economic development and completed the productivist reduction of 'active ageing'.

> Member States should develop active ageing policies by adopting measures to maintain working capacity and skills of older workers, to introduce flexible working arrangements, and to raise employers' awareness of older workers' potential. They should ensure that older workers have sufficient access to further education and training and review tax and social protection systems with the aim of removing disincentives and creating incentives for them to remain in the labour market. (European Commission 2001, p. 50)

Although there have been periodic nods towards the more comprehensive notion of active ageing espoused by the WHO, and EC in 1999, this narrow focus on employment and extending working lives continues to dominate EU policymaking. This is not to undervalue the importance of promoting employment opportunities for older workers and combating age barriers but this is not active ageing (Walker 2009; Walker and Maltby 2012).

So what happened? In a nutshell, economic and market issues dominate European policymaking and are invariably prioritised. Moreover, while the aim of extending working lives is not incompatible with active ageing, the policy instruments at EU level do not lend themselves to the implementation of a more comprehensive version of this concept. Perhaps also the wide range of actions implied by a comprehensive approach are just too daunting for European policymakers to contemplate. Maybe the problem is the same as at the national level, where policymakers' horizons are set by the short-term electoral cycle, and active ageing is a long-term strategy. Then there is the possibility that they simply do not *get* active ageing, that is, they do not grasp it fully and therefore are not aware of its enormous potential.

The Promise of Active Ageing

To understand the potential of active ageing and its reach far beyond the labour market, it is necessary to go back to the original concept advanced by the WHO (2002, p. 12):

> Active ageing is the process of optimising opportunities for health, participation and security in order to enhance quality of life as people age.

The crucial phrase in this definition is 'as people age' rather than 'when they are old'. This recognises that ageing is lifelong and results from a lifetime of cumulative wear and tear, rather than being a phenomenon that commences at or around pension ages. The great promise of active ageing, in social and economic terms, hangs on the fact demonstrated in Chap. 7 that in the lifelong process of ageing the external environmental determinants are far more influential than the genetic ones, by as much as four or five to one (Gems and Partridge 2013). There is wide biogerontological consensus on this point (López-Otin et al. 2013) and also on the fact that there is no such thing as a gene for ageing—the ideas that human beings are programmed to live a fixed period or to age in a certain way (Kirkwood 2005). The key environmental risk factors include both political economy variables such as deprivation, low socio-economic status, air pollution, unhealthy food production and occupational stress, and behavioural ones such as smoking, lack of exercise, and poor diet. For example, there is clear evidence, derived from longitudinal data analyses, of associations between low socio-economic status and area deprivation in childhood and lower levels of functioning in old age (Walker 2018; see also http://www.newdynamics.group.shef.ac.uk/).

The pre-eminence of the external, non-genetic, factors in human ageing provides the key to realising the potential of active ageing as well as pointing to the sorts of policy measures required to do so. The risk factors that inflict damage on the human body and mind and often result in the chronic conditions associated with loss of function in later life, or biological ageing, can be moderated or prevented entirely. These chronic conditions, often experienced as multi-morbidities, such as coronary heart disease, stroke, type-2 diabetes and so on, may cut short lives prematurely or result in disabilities which reduce functioning and quality of life, and which may require treatment or care. But, adopting a life course approach to the aetiology of these chronic conditions highlights the potential to reduce and postpone them. This is the primary aim of the active ageing approach and why its adoption is so important for ageing societies.

As demonstrated throughout this book, the scientific evidence supporting active ageing is robust and growing continuously (http://mopact.group.shef.ac.uk; Fernandez-Ballesteros et al. 2013). Key examples include the beneficial effects of physical exercise: the active had 33–50 per cent lower risk of developing type-2 diabetes than the inactive; and the moderately active have a 20 per cent lower risk of stroke incidence than the inactive. With respect to diet, the association between fruit and vegetable intake and the reduced risk of cardio-vascular disease and cancer is very clear: reductions in the risk of cardiovascular disease and all causes of mortality are observed in an intake of 800 g/day of both and, for all causes of cancer, 600 g/day (Aune et al. 2017). As these examples, from many, illustrate it is not scientific evidence that is lacking but the purposeful translation of this evidence into practical interventions and policy strategies. On this matter, it may be that the tendency of policymakers to overemphasise employment has actually harmed the cause of a more comprehensive approach.

Realising the Promise

What would a comprehensive strategy on active ageing look like? First of all it would emphasise *prevention*. Implicit within this priority is a life course perspective because it is essential to intervene as early as possible, including during pregnancy and early life, to prevent later life chronic morbidities. Mid-life interventions are critical too because, between the ages of 35 and 55 there is a sevenfold increase in the prevalence of multi-morbidity. While interventions in later life, including a new generation of drugs (geroprotectors), can also be beneficial, it is essential also to activate secondary prevention and to ensure that remedial supports are available. A similar life course preventative approach should apply to working life. Here, rather than focussing only on late working life, policymakers should encourage lifelong education and continuous training to prevent skill loss.

Second, the 'active' in active ageing should encompass all activities that contribute to physical and mental health and not just paid employment. Indeed, the myopic policy focus on work may conflict with other personally and socially rewarding activities, such as volunteering. Third, active ageing should be inclusive of all older people, even those with frailty. This is because of the danger of excluding the old-old and because the link between activity and health (including mental stimulation) holds firm into advanced old age. Fourth, an active ageing strategy should respect diversity and not take the 'one size fits all' approach of the working longer policy. For example, gender sensitivity is critical (Foster and Walker 2013). Cultural and national diversities must be taken into account, for example, variations in cultural norms and institutional development between the north and south, and east and west in Europe. Finally, the idea of active ageing embodies both rights and obligations: the rights to social protection, lifelong education and training, and so on should be accompanied by obligations to take advantage of these opportunities and to remain active in other ways. From a gender perspective, the realisation of such obligations requires specific supportive measures to enable women to participate on an equal basis to men. This is not to suggest that sanctions should be applied for non-compliance but, rather, that the redefinition of the social contract that this comprehensive form of active ageing implies, would entail education from an early age about the need for personal as well as socio-political responsibility in the adoption of prevention as a guideline for life.

It is clear from the above that realising a comprehensive strategy on active ageing would require a partnership between citizens and society. In this partnership, the role of the state is to enable, facilitate, and motivate citizens and to provide high-quality social protection. It also has a critical role in regulating the meso sectors, especially employers, to ensure healthy working conditions and continuous training. For their part, citizens have a duty to take advantage of lifelong learning and continuous training opportunities and to promote their own lifelong health and well-being. The big policy challenge is to join-up the separate domains—employment, health, social protection, social inclusion, education, transport, and so on—so that they are mutually supportive. Unfortunately, this is not likely to happen when the EU and national policy machines are focused primarily on paid employment.

Distilling drastically all that has gone before the five prerequisites for the realisation of active ageing in Europe are:

- First, a new more optimistic paradigm of ageing needs to be taken onboard by all key policy actors and promoted; this includes the destruction of the old worn-out assumptions and stereotypes of what 'old age' is and their replacement by a more dynamic perspective on ageing as a whole and on older people as agents of change with social and economic rights.
- Second, a life course perspective is essential, putting a strong emphasis on active ageing at all ages and joining-up all the relevant policy domains.

- Third, the heterogeneity of older people (aged 50 and over) needs to be appreciated. Thus, active ageing strategies have to be made relevant for everyone and not reduced to elites who are better educated and enjoy good health in later life.
- Fourth, the multi-dimensional nature of active ageing must capture all aspects of social and economic activities and not be reduced simply to longer working lives—there is more to life than work.
- Fifth, top-down active ageing policies will not work in Europe because of the wide variations in institutional capacities and enabling environments. These must be recognised and tailor-made strategies developed if active ageing is to be possible in all Member States.

References

Aune, D., et al. (2017). Fruit and Vegetable Intake and the Risk of Cardiovascular Disease, Total Cancer and All-cause Mortality. *International Journal of Epidemiology*, 1–28. https://doi.org/10.1093/ije/dyw319.

European Commission. (1999). *Towards a Europe for All Ages*. Brussels: EC.

European Commission. (2001). *Guidelines for Member States' Employment Policies for the Year 2002*. Brussels: EC.

Fernandez-Ballesteros, R., Robine, J. M., Walker, A., & Kalache, A. (eds) (2013). 'Active Ageing', Special Issue of Current Gerontology and Geriatrics Research.

Foster, L., & Walker, A. (2013). Gender and Active Ageing in Europe. *European Journal of Ageing, 10*(1), 3–10.

Gems, D., & Partridge, L. (2013). Genetics of Longevity in Model Organisms: Debate and Paradigm Shifts. *Annual Review of Physiology, 75*, 621–644.

Kirkwood, T. (2005). Understanding the Odd Science of Ageing. *Cell, 120*, 437–447.

Lopez-Otin, C., et al. (2013). The Hallmarks of Cell Aging. *Cell, 153*, 1194–1217.

UN. (2002). *Report of the Second World Assembly on Ageing*. New York: United Nations.

Walker, A. (2009). The Emergence and Application of Active Ageing in Europe. *Journal of Aging and Social Policy, 21*, 75–93.

Walker, A. (2018). Why the UK Needs a Social Policy on Ageing. *Journal of Social Policy, 47*(2), 253–273.

Walker, A., & Maltby, T. (2012). Active Ageing: A Strategic Policy Solution to Demographic Ageing in the European Union. *International Journal of Social Welfare, 21*(s1), S117–S130.

WHO. (2002). *Active Ageing: A Policy Framework*. Geneva: WHO.

FSC
www.fsc.org
MIX
Papier | Fördert
gute Waldnutzung
FSC® C083411

Kolibri 360 GmbH
Ferdinand-Jühlke-Straße 7
99095 Erfurt, Deutschland
produktsicherheit@kolibri360.de